Lecture Notes in Physics

348

A. Bohm
M. Gadella

Dirac Kets, Gamow Vectors and Gel'fand Triplets

The Rigged Hilbert Space Formulation of Quantum Mechanics

Lectures in Mathematical Physics at the University of Texas at Austin

Edited by A. Bohm and J. D. Dollard

Springer-Verlag
Berlin Heidelberg GmbH

Authors

Arno Bohm
Department of Physics, The University of Texas
Austin, TX 78712, USA

Manuel Gadella
Facultad de Ciencias, Universidad de Valladolid
E-47011 Valladolid, Spain

ISBN 978-3-662-13751-2 ISBN 978-3-540-46859-2 (eBook)
DOI 10.1007/978-3-540-46859-2

Originally published by Springer-Verlag Berlin Heidelberg New York in 1989
Softcover reprint of the hardcover 1st edition 1989

2153/3140-543210 – Printed on acid-free paper

TABLE OF CONTENTS

CHAPTER I

CHAPTER II

CHAPTER III

CHAPTER IV

PREFACE

This book is an expansion of a little booklet on rigged Hilbert spaces an quantum mechanics[1]. The purpose of that book was to present a mathematical foundation of Dirac's bracket formalism.

When Dirac introduced his formalism[2], the mathematics for it did not exist and von Neumann's Hilbert space formulation[3] became the mathematically rigorous formulation of quantum mechanics. Physicists preferred the Dirac formalism because it had many advantages: the observables formed an algebra on the whole space of physical states and could be treated as continuous operators; every hermitian observable had a complete set of eigenkets and the set of their "eigenvalues" could be discrete or continuous; the wave functions were well-behaved smooth functions and to each state vector corresponded one wave function (not a set of functions which can vary wildly on a set of Lebesgue measure zero). Most importantly, using the Dirac formalism did not lead to any disasters in physical calculations. From this, one concludes that the Dirac's formalisme was fine, only the right mathematics had to be created for it.

This mathematics was developed in two steps. Inspired by the δ-"fnction", which was a product of Dirac's formalism, the thoery of distributions was developed and for the first time systematically presented by L. Schwartz[4] (1950-1951). On this basis, I. M. Gel'fand and collaborators[5] introduced the rigged Hilbert space (also called Gel'fand triple) in connection with the spectral theory of self-adjoint operators, and proved, together with Maurin[6], the nuclear spectral theorem (1955-1959).

Therewith, enough mathematics had been created for the rigged Hilbert space formulation of quantum mechanics, which was introduced around 1965 by A. Bohm and J. E. Roberts[7]. Its goal was not to give a new physical description but to make sense of Dirac's heuristic formalism, which is the basis of the algebraic method. A few years ago another mathematical interpretation of Dirac's formalism was given which avoids the sharp distinction between well-behaved state vectors and generalized eigenvectors[8]. But, in the meantime, the rigged Hilbert space formulation has led to something new which was not anticipated at the time of its introduction: the Gamow vectors for decaying states.

The rigged Hilbert space formulation is exemplary for the role that rigorous mathematics play in physics: making sense of heuristic ideas not asserting that they are nonsense. Dirac's formalism of bras and kets is the heuristic idea. Von Neumann's Hilbert space formulation asserted that it is mathematical nonsense. The rigged Hilbert space is the rigorous mathematics that makes sense of Dirac's formalism. But just making sense of heuristic ideas is not were the role of rigorous mathematics ends. After a new mathematical language has been learned in physics, it is used to think and to speak. In this way one will be led to new ideas and deeper insights which could not be attained with the old language. This also happened with the rigged Hilbert space formulation. The new mathematical language contained entities that can describe exponentially decaying states, which we have called Gamow vectors. This was not foreseen when the rigged Hilbert space formulation of quantum mechanics was conceived. These new developments are presented in the second part of the book (Chaps. II-IV).

The rigged Hilbert space (Gel'fand triplet) is one of the most beautiful creations of mathematics. As it is our intention to make this little-known area of mathematics accessible to a wider audience of physical scientists and mathematical physicists, we have limited ourselves to a simplified picture. The whole subject is developed in terms of the simplests examples (the oscillator in the first part and a single pole resonance in the second) and the mathematical notions have been introduced in the least general form, which is compatible with the purpose of this book. We hope that this simplified picture still conveys enough of the beauty of this mathematical creation to make it not only accessible, but also attractive to physical scientists. And we hope that the physical examples are elementary enough to give mathematicians an idea of what their creations are good for, without burdening them with too many physical details. The general mathematics has been treated in several monographs[5]; the physical applications can be found in books of quantum physics[9].

1.) A. Bohm. The Rigged Hilbert Space and Quantum Mechanics. Springer Lecture Notes in Physics, vol. 78 (1978).

2.) P. A. M. Dirac. The Principles of Quantum Mechanics. 4th edition. Clarendon Press. Oxford 1958, and earlier editions.

3.) J. von Neumann. Mathematical Foundations of Quantum Mechanics. Springer. Berlin 1932. Princeton University Press 1955.

4.) L. Schwartz. Theorie des Distibutions. Hermann. Paris 1973, and references thereof.

5.) I. M. Gel'fand and G. P. Shilov. Generalized Functions. vol 4. Academic Press. New York 1964, and references thereof.

6.) K. Maurin. General Eigenfunction Expansions and Unitary Representations of Topological Groups. Polish Scientific Publishers. Warsaw 1968, and references thereof.

7.) J. E. Roberts, J. Math. Phys., 7, 1097, (1966); Commun. Math. Phys., 3, 98, (1966). A. Bohm, "The Rigged Hilbert Space in Quantum Physics", ICTP Report No. 4, Trieste (1965); Boulder Lectures in Theoretical Physics (A. O. Barut, editor), vol 9A (1966).

8.) S. J. L. van Eijndhoven and J. de Graaf. A Mathematical Introduction to Dirac´s Formalism. North Holland. Amsterdam 1986.

9.) A. Bohm. Quantum Mechanics: Foundations and Applications. 2nd Edition. Springer. Berlin, Heidelberg 1986.

Acknowledgements

This book was started while both authors were visiting the University of Wuerzburg. We gratefully acknowledge the advice and the hospitality offered by Karl Kraus and partial support from the Alexander Humboldt Foundation. This project was aditionally supported by a NATO grant for international collaboration in research. We also wish to express our gratitude to I. M. Gel'fand for fruitful discussions.

CHAPTER I

INTRODUCTION

To obtain a physical theory means to obtain a mathematical image of a physical system. The mathematical image for the domain of quantum phenomena is, according to von Neumann[1]) the set of operators in a Hilbert space. For the case that no superselection rules are present (irreducible systems) the physical observables are identified with the set of all self-adjoint operators in the Hilbert space H and the ("pure") physical states are identified with the set of all unit rays, or equivalently, the one dimensional projection operators in H.

This one-to-one correspondence between the physical objects of a system and the operators in a Hilbert space has two deficiencies:

1) It does not contain all the information one has in quantum mechanics.
2) It contains more information than one can ever obtain in physics.

To elaborate on 1). It is well known that all (separable infinite dimensional) Hilbert spaces are isomorphic and so are-rougly speaking-their algebras of operators. This means that all physical systems would be equivalent, which is obviously not the case (unless one has only one physical system, which then would have to be the microphysical world).

In every practical case one has to consider an isolated physical system which -though it has to be large enough so that it does not become trivial- must be small enough that one may command a view of its mathematical image.

The various microphysical systems in nature have quite distinct properties and must therefore have quite distnct mathematical images. Or, in other words, for a particular system some observables are completely unphysical, e.g., if a system consists of the bound states of the hydrogen atom, then for this physical system one can never prepare a state in which the electron position is in a narrow domain of the space, because a measurement of the observable position will destroy this physical system.

Thus for a particular physical system not all mathematical observables are physical observables.

Further, for a particular physical system some observables appear to be more basic than others, and which observables are more basic depends upon the particular physical system.

Therefore, it appears appropiate to make a distnction between observables and operators.

An observable is a physical quantity and the physical definition of an observable consists of either a prescription for measuring the quantity itself or a definite expression for it in terms of other measurable quantities.

If not all operators are observables then also not all vectors can be physical states, because the preparation of a physical state constitutes the measurement of one or more observables, and a pure state that results from this measurement is an eigenstate of the observables. Thus eigenstates of operators which are not observables cannot be prepared.

Further -and this is connected with 2)- there are vectors which are connected with operators that are observables but which are not physically realizable. These are the vectors that lie outside the domain of definition of an observable that is represented by an unbounded operator, e.g., the vector which would correspond to the state of infinite energy.

With regard to 2): The Hilbert space is infinite dimensional and it is infinite dimensional in a very particular sense, namely it is complete with respect to a very particular topology, i.e., with respect to a very particular meaning of convergence of infinite sequences. Since an infinite number of states can never be prepared, physical measurements cannot tell us anything about infinite sequences, but can at most give us information about arbitrarily large but finite sequences. Therefore physics cannot give us sufficient information to tell how to take the limit to infinite,i.e., how to choose the topology. All that physics can tell us is that we need a linear space sometimes finite but in general, of arbitrarily large dimensions.

We believe that the choice of the topology will always be a mathematical generalization, but then one should choose the topology such that it is most convenient.

At the time when quantum mechanics was developed the Hilbert space was the only linear topological space that had strutures which were required by quantum mechanics. That is, it is linear to fulfill the superposition principle, and it could accomodate infinite dimensional matrices and differential operators. Therefore, it was natural that von Neumann chose the Hilbert space when he wanted to give a mathematically rigorous formulation of quantum mechanics. With this choice, though, he could not accomodate all the features of Dirac´s formulation[2)] of quantum mechanics, which von Neumann says is "scarcely surpassed in brevity and elegance" but "in no way satisfies the requirements of mathematical rigour".

Since that time , stimulated by the Dirac formalism, and in particular by the ill defined notion of Dirac´s delta, a new branch of mathematics, the theory of distributions, has been developed. The abstraction of this theory provided the mathematical tool for a mathematically rigorous formulation of quantum mechanics, which not only overcomes the above mentioned shortcomings of von Neumann´s axioms but also has all the niceties of the Dirac formalism.

This mathematical tool is the Rigged Hilbert Space, which was introduced around 1960 by Gel´fand[3)] and collaborators and Maurin[4)] in connection with the spectral theory of self-adjoint operators. Its use for the mathematical foundation of Quantum Mechanics was suggested around 1965[5,6)], the suggestion to choose the topology such that the operators are continuous (for the canonical commutation relation) was made slightly earlier.[7)]

We formulate the general framework by replacing the von Neumann axiom which states the one-to-one correspondence between observables and self-adjoint operators in the Hilbert space by the following basic assumption.[8)]

A physical observable is represented by a linear operator in a linear space (space of states).

The mathematical image of a physical system is an algebra A of linear operators on a linear scalar product space. The linear space is equipped with a nuclear topology that makes this algebra an algebra of continuous operators. Also, the operator $\Delta = \Sigma X^2{}_i$, where X_i, i =1,2,... are the generators of *A* is essentially self-adjoint (e.s.a.).

We emphasize again: An observable is the physical quantity which is defined by the prescription for its measurement. The algebra is the mathematical structure which is defined by the algebraic relations between its generators which represent some basic physical quantities and by other mathematical conditions, like the space in which the elements of the algebra act as operators.

The connection between the quantities calculated in the mathematical image, such as the matrix elements $(\psi,A\psi)$ or Tr(AW) and the quantities measured in the physical world, such as the expectation value (average value of a measurement) M(A), is given by $M(A) = (\psi,A\psi)$ or in general $M(A) = Tr(AW)$. These connections are stated by further basic assumptions which we will not discuss in this paper.

Within the framework of the above stated basic assumption the actual physical question is how to find the algebra *A* . In non-relativistic quantum mechanics one can usually obtain this algebra from correspondence with the corresponding classical system. When no corresponding classical system exists one has to conjecture this algebra from the experimental data.

I. THE ALGEBRAIC STRUCTURE OF THE SPACE OF STATES

We shall treat in detail one of the simplest physical models, the one-dimensional harmonic oscillator, in the framework of the basic assumption stated in the preceding section. We will give a formulation which easily generalizes to more complicated physical models, and shall remark for which algebras these generalizations are already known.[9]

The algebra[10] A of operators for the one dimensional harmonic oscillator is generated by the operators:

H representing the observable energy.

P representing the observable momentum.

Q representing the observable position.

The defining algebraic relations are:

$$PQ - QP = -i\hbar I \qquad (1\text{-}1)$$

$$H = \frac{1}{2m}P^2 + \frac{m\omega^2}{2}Q^2$$

$\hbar$, m,ω are constants, universal or characteristic of the system.

A is an algebra of linear operators on a linear space, Φ, with scalar product [11] (.,.) (but Φ is not a Hilbert space). Further, P, Q, H are symmetric operators, i.e.

$$(A\varphi,\psi) = (\varphi,A\psi), \quad A= P,Q \text{ or } H. \qquad (1\text{-}2)$$

(1-1) and (1-2) do not specify the mathematical structure completely. There are many such linear spaces in which A is an algebra of operators (whether the requirements that its topology be nuclear and even the weakest puts sufficient restrictions is unknown). Therefore we have to make one further assumption to specify the algebra of operators, i.e., the representation of (1-1) and (1-2). This requirement can be formulated in two equivalent ways:

H is essentially self-adjoint[12] (e.s.a.). (1-3-a)

or

There exists at least one <u>eigenvector</u> of H. (1-3-b)

We will use this requirement in the form (1-3-b), as it is this form in which it is always used by physicists.

<u>Remark</u>. We remark that either of these requirements leads to a representation of (1-1) which integrates to a representation of the group generated by P, Q, I (Weyl group); (1-3-a) by a therem of Nelson[13] and (1-3-b) because it leads to the well known ladder representation and ladder representations are always integrable[14] (in this particular case one can easily see that H in the ladder representation will be e.s.a.)

The generalization of (1-3) is the requirement that $\Delta = \Sigma X^2_i$ be e.s.a.. For the case that A is the enveloping algebra of a Lie group, this will always lead to an integrable representation (Nelson theorem[15]), i.e., a representation that is connected with a representation of a group (even though no symmetry may be involved).

The procedure to find the ladder representation is well known and will be sketched only very briefly. One defines.

$$a = \frac{1}{\sqrt{2}} \sqrt{\frac{m\omega}{\hbar}}\, Q + \frac{i}{\sqrt{m\omega\hbar}}\, P$$

$$a^{+} = \frac{1}{\sqrt{2}} \sqrt{\frac{m\omega}{\hbar}}\, Q - \frac{i}{\sqrt{m\omega\hbar}}\, P$$

(1-4)

$$N = a^{+} a = \frac{1}{\omega\hbar} H - \frac{1}{2} I$$

(1-5)

These operators clearly fulfill

$(\varphi, a\psi) = (a^{+}\varphi, \psi)$ for every $\varphi, \psi \in \Phi$

and

$(\varphi, N\psi) = (N\varphi, \psi)$ for every $\varphi, \psi \in \Phi$

as a consequence of (1-2).

As a consequence of (1-1), a and a^{+} fulfill

$a\, a^{+} - a^{+} a = I$ (1-6)

As a consequence of (1-3-a), N es e.s.a. (1-7-a)

As a consequence of (1-3-b), there exists φ_λ such that

$$N\varphi_\lambda = \lambda\varphi_\lambda \tag{1-7-b}$$

From the commutation relations (1-6), then follows

$$N(a\varphi_\lambda) = a^{+} a\, a\, \varphi_\lambda = (a\, a^{+} - I)\, a\varphi_\lambda = a(a^{+} a - I)\varphi_\lambda$$

$$= a(N - I)\varphi_\lambda = a(\lambda - 1)\varphi_\lambda = (\lambda - 1)a\varphi_\lambda \tag{1–8}$$

i.e. $a\varphi_\lambda$ is an eigenvector with eigenvalue $(\lambda-1)$ or $a\varphi_\lambda = 0$. Further from c.r. (1-6) follows

$$\|a^{+}\varphi_\lambda\|^2 = (\varphi_\lambda, a^{+}a\varphi_\lambda) + (\varphi_\lambda, I\varphi_\lambda)$$

$$= \|a\varphi_\lambda\|^2 + \|\varphi_\lambda\|^2 \neq 0 \text{ as } \varphi_\lambda \neq 0 \tag{1-9}$$

i.e. $a^+\varphi_\lambda \neq 0$ always.

Further, it follows from (1-6)

$$N(a^+\varphi_\lambda) = (\lambda-1)a^+\varphi_\lambda \tag{1-10}$$

i.e. $a^+\varphi_\lambda$ is always an eigenvector of N with eigenvalue $(\lambda + 1)$. Starting with φ_λ, which was assumed to exist, one can obtain eigenvectors

$$\varphi_{\lambda-m} = a^m \varphi_\lambda \qquad m = 0,1,2,...$$

with eigenvalue $(\lambda-m)$, of N.

$$N\varphi_{\lambda-m} = (\lambda-m)\, \varphi_{\lambda-m}.$$

After a finite number of steps

$$a^m\varphi_\lambda = 0$$

Proof:

$$(\varphi_{\lambda-m}, N\, \varphi_{\lambda-m}) = (\lambda-m)\varphi_{\lambda-m}, \varphi_{\lambda-m}) = (\varphi_{\lambda-m}, a^+a\varphi_{\lambda-m}) = ||a\varphi_{\lambda-m}||^2$$

therefore

$$(\lambda-m) = \frac{||a\varphi_{\lambda-m}||^2}{||\varphi_{\lambda-m}||^2} \geq 0 \tag{1-11}$$

i.e.,there exists a $\varphi_{\lambda-m}$ such that

$$a\varphi_{\lambda-m} = 0$$

We call φ_0 the normalized vector for which $a\varphi_0 = 0$ i.e.

$$\varphi_0 = \frac{\varphi_0}{||\varphi_0||} \; ; N\varphi_0 = 0 \tag{1-12}$$

Then one define the states.

$$\varphi_0 \tag{1-13}$$

$$\varphi_1 = \frac{1}{\sqrt{1!}} a^+ \varphi_0$$

$$\varphi_2 = \frac{1}{\sqrt{2!}} (a^+)^2 \varphi_0$$

....

....

....

$$\varphi_n = \frac{1}{\sqrt{n!}} (a^+)^n \varphi_0$$

which have the property.

$$1)\ \|\varphi_n\| = 1\ ;\ 2)\ N\varphi_n = n\varphi_n\ ;\ 3)\ (\varphi_n, \varphi_m) = \delta_{nm} \tag{1-14}$$

4) For every φ_n there exists $\varphi_{n+1} \neq 0$

The linear space spanned by φ_0, φ_1, ..., φ_n,... we call Ψ, i.e. Ψ is the set of all vectors

$$\psi = \sum_{n=0,1...}^{m} \alpha^{(n)} \varphi_n \tag{1-15}$$

where $\alpha^{(n)} \in C$ is the set of complex numbers and m is a natural number which is arbitrarily large but finite. We call R_1 the space spanned by φ_1, i.e.

$$R_i = \{\alpha\varphi_i | \alpha \in C\},\ (R_i \text{ are the energy eigenspaces})$$

Using R_i instead of φ_i we have a formulation which immediately applies when R_i is not one dimensional. Then Ψ is the algebraic direct sum of the R_i.

$$\Psi = \Sigma R_i \tag{1-16}$$

In other words Ψ is the set of all sequences $\varphi = (\varphi_0, \varphi_1, ...\varphi_n, 0, 0, ..)$ with $\varphi_i \in R_i$ where the algebraic operations of the linear space are defined by

$$\varphi + \psi = (\varphi_0 + \psi_0, \varphi_1 + \psi_1, ...) \tag{1-17}$$

$$\alpha\varphi = (\alpha\varphi_0, \alpha\varphi_1,...)$$

The scalar product is

$$(\varphi,\psi) = \sum_{i=0}^{m} (\varphi_i, \psi_i) \tag{1-18}$$

And the norm is

$$||\varphi||^2 = \sum_{i=0}^{m} (\varphi_i,\varphi_i) \tag{1-19}$$

As the sum goes only up to an arbitrary large but finite number m, the question of convergence of the sum of numbers (φ_i,φ_i) does not arise.

Ψ is a linear space without any topological structure. We now equip Ψ with a topology, i.e., we construct a linear topological space.

II. THE TOPOLOGICAL STRUCTURE OF THE SPACE OF STATES

A set Φ is a linear topological space (in which the first axiom of countability is fulfilled) if: a) Φ is a linear space; b) for a sequence of elements of Φ the notion of a limit element is defined (the limit is unique and a subsequence of a convergent sequence converges to the same point); and c) the algebraic operations of the linear space are continuous i.e.

1a) If $\Phi \ni \varphi_n \to \varphi \in \Phi$ then also $\alpha\varphi_n \to \alpha\varphi$ for $\alpha \in C$

1b) If $C \ni \alpha_n \to \alpha \in C$ then also $\alpha_n\varphi \to \alpha\varphi$ for every $\varphi \in \Phi$ (1–20)

2) If $\varphi_n \to \varphi$ and $\psi_n \to \psi$ then $\varphi_n + \psi_n \to \varphi + \psi$ for $\varphi_n, \psi_n, \varphi, \psi \in \Phi$; n=1,2,...

This is not the most general definition of a linear topological space but it is sufficient for our present purpose.

In a given linear space the convergence can be defined in various ways leading to various linear topological spaces. All the topological notions like continuity, denseness, boundedness, closure, completeness... depend then on the definition of convergence.

We shall introduce into Ψ two different topologies. As a first example we introduce into Ψ the well known Hilbert space topology τ_H . Definition[16]:

$$\varphi_\gamma \to \varphi \text{ for } \gamma \to \infty \Leftrightarrow \|\varphi_\gamma - \varphi\| \to 0 \text{ for } \gamma \to \infty.$$

To understand the relation between Ψ and the Hilbert space H , let us introduce the following notion: A sequence $\{x_n\}$ of elements in a space with a norm (normed space) is called a Cauchy sequence with respect to τ_H if for every $\varepsilon > 0$ there exists an N such that

$$\|x_n - x_m\| < \varepsilon \tag{1–21}$$

for every n,m > N. <u>Definition</u>: A linear topological space is called complete if every Cauchy sequence $\{x_n\}$ has a limit x which is an element of the space R.

If a space is not complete it can be completed by adjoining all limit elements of Cauchy sequences to it .

It is easy tosee that Ψ is not τ_H complete. Let us consider the infinite sequence

$$h = (h_1, h_2, \ldots, h_n, \ldots) \text{ with } h_i \in R_i \tag{1-22}$$

$h_i \neq 0$ for all i

$$\text{and} \sum_{i=0}^{\infty} ||h_i||^2 < \infty$$

From this follows

$$||h_n|| \to 0 \quad \text{for } n \to \infty \text{ and} \tag{1-23}$$

$$\sum_{i=n+1}^{\infty} ||h_i||^2 \to 0 \ \text{ for } n \to \infty$$

h is <u>not</u> an element of Ψ because it is an infinite sequence.

Let us consider the sequence

$s_1, s_2, \ldots$ with $s_i \in \Psi$

where

$$s_n = (h_1, h_2, \ldots, h_n, 0, 0, \ldots) \tag{1-24}$$

As $s_n - s_m = (0,0,\ldots 0, h_{n+1}, h_{n+2}, \ldots, h_m, 0, 0, \ldots)$ we have

$$||s_n - s_m|| = ||h_{n+1} + h_{n+2} + \ldots + h_m|| \to 0 \text{ for } n \to \infty \tag{1-25}$$

i.e. $\{s_n\}$ is a Cauchy sequence.

The τ_H limit element of s_n is h, because

$$||s_n - h|| = \sum_{i=n+1}^{\infty} ||h_i||^2 \to 0 \ \text{ for } n \to \infty \tag{1-26}$$

But $h \notin \Psi$.

If we adjoin to Ψ all the limits of Cauchy sequences, i.e. take in addition to the elements $\varphi = (\varphi_1, \varphi_2, \ldots, \varphi_n, 0, 0, \ldots)$, n arbitrarily large, the elements h of the kind defined in (1-22), then we obtain a complete normed space in which the norm is given by the scalar product, this is the Hilbert space H. Thus, the Hilbert space H is the completion [17] of Ψ with respect to the topology defined by (1-20).

<u>Definition</u>: A subspace S of a complete topological space X is called dense if

$$S \cup \{\text{all limit points of } S\} = X \tag{1-27}$$

Thus Ψ is a τ_H -dense subspace of H .

The space H is the space of all $h = (h_0, h_1, \ldots)$, $h_i \in R_i$ which fulfill (1-22); this space is written as:

$$H = \sum_{i=0}^{\infty} \oplus R_i \tag{1-28}$$

and is called the Hilbertian direct sum or orthogonal direct sum of the R_i.

Thus, the Hilbertian direct sum is the completion of the algebraic direct sum with respect to τ_H.

We now introduce into Ψ a new topology which we call τ_Φ . τ_Φ is defined in the following way:

We take the original scalar product (.,.) and define the following quantities

$$(\varphi,\psi)_p = (\varphi,(N+1)^p \psi), \quad p = 0,1,2,\ldots \text{ and } \|\varphi\|_p = (\varphi,\varphi)_p \tag{1-29}$$

Remark: Before proceeding, we note some properties of the operator N. N is essentially self-adjoint. This is easily proved using one of the criteria of essentially self-adjointness. Lemma: An operator A is e.s.a. if $(A + I)^{-1}$ is continuous and has a dense domain on H .

The spectrum of $(N + I)^{-1}$ is $(n+1)^{-1}$, consequently it is continuous. Its domain is Ψ, which is dense in H . Consequently, N is e.s.a.. As a consequence N + I is e.s.a.. Further $(N + I)^p$ for every p is e.s.a.[18].

From the spectrum of $(N + I)^p$ one sees immediatly that $(N + I)^p$ is positive definite. From this it is easy to see that $(\varphi,\psi)_p$ fulfills the condition of a scalar product and

$$\|\varphi\|_0 \le \|\varphi\|_1 \le \|\varphi\|_2 \le \ldots \tag{1–30}$$

Remark: Further, these norms are compatible, i.e. if a sequence converges with respect to one norm and is a Cauchy sequence with respect to another, then it also converges with respect to this other norm.

A space in which a countable number of norms (scalar products) are defined is called a countably normed (countably scalar product) space. We now define τ_Φ. Definition

$$\varphi_\gamma \to \varphi \text{ with respect to } \tau_\Phi \Leftrightarrow \|\varphi_\gamma - \varphi\|_p \to 0 \text{ for every p} \tag{1-31}$$

From this definition one immediately sees

From $\varphi_\gamma \to 0$ with respect to τ_Φ $\Rightarrow$ follows $\varphi_\gamma \to 0$ with respect to τ_H (1-32)

but not vice versa. Therefore τ_Φ is called stronger (finer) than τ_H and τ_H is called weaker (coarser) than τ_Φ. Definition: A sequence $\{x_\gamma\}$ is called a τ_Φ sequence if for every p and for every $\varepsilon > 0$ there exists an $N = N(\varepsilon,p)$ such that

$$||x_\gamma - x_\mu||_p < \varepsilon \text{ for every } \mu,\gamma > N \tag{1-33}$$

There are more τ_Φ –Cauchy sequences than τ_Φ –Cauchy sequences because for τ_Φ –Cauchy sequences (1-33) needs to be fulfilled only for p = 0.

We complete[17)] Ψ with respect to τ_Φ i.e. adjoin to Ψ the limit points of all τ_Φ -Cauchy sequences. The linear topological space that we obtain by completing Ψ with respect to τ_Φ we call Φ. Then

$$\Psi \subset \Phi. \quad \Psi \text{ is } \tau_\Phi\text{- dense in } \Phi$$

Φ is called a countably Hilbert space. As there are more τ_H -Cauchy sequences than τ_Φ-Cauchy sequences, it must be that

$$\Psi \subset \Phi \subset H \tag{1-34}$$

Thus, Ψ is τ_Φ-dense in Φ; Ψ is τ_H -dense in H , and Φ is τ_H -dense in H as already Ψ is τ_H -dense in H .

To get a feeling for Φ let us see which of the infinite sequences that are elements of H are elements of Φ. In order that $h = (h_0,h_1,...) \in H$ be a limit point of a τ_Φ –Cauchy sequence $\{s_n\}$ with $s_n = (h_1,h_2,...,h_n,0,0,....) \in \Psi$, i.e., in order that: $s_n \to h$, with respect to τ_Φ, one must have:

$$||s_n - h||_p \to 0 \text{ for } \underline{\text{every}} \text{ p,i.e.} \tag{1-35}$$

$$((s_n - h), (N + I)^p(s_n - h)) \to 0 \text{ for every p} \tag{1-36}$$

i.e., one must have

$$\sum_{i=n+1}^{\infty} (h_i, (N+I)^p h_i) = \sum_{i=n+1}^{\infty} (i+1)^p \, \|h_i\|^2 \to 0$$

(1-37)

for every p. Therefore, only those $h \in H$ are τ_Φ–limit points of τ_Φ-Cauchy sequences of elements of Ψ- that means elements of Ψ- for which

$$\sum_{i=1}^{\infty} (i+1)^p \, \|h_i\|^2 < \infty$$

(1-38)

Φ is called the τ_Φ direct sum of R_n and is written

$$\Phi = \sum_{n=0}^{\infty} \oplus R_n$$

(1-39)

To see what the use of the various spaces might mean for physics we recall that R_n are the energy eigenspaces with energy value

$$E_n = \hbar \, \omega \, (n + \frac{1}{2}).$$

(1-40)

Thus, Ψ describes states with arbitrarily high but finite energy. H contains the "infinite energy" state, and Φ contains states which have an "infinitely small admixture of infinite energy states".

Clearly physics can only tell us something about Ψ. In fact for most real physical systems. e.g., diatomic molecules, whose idealization is the harmonic oscillator, only the very lowest energy levels are relevant; for higher energy the diatomic molecule is no longer a harmonic oscillator and finally not even an oscillator. H and Φ appears "closer" to reality.

The question there is, why do we want these mathematical idealizations and why do we have a preference for one over the other?.

The advantage that Φ has over H is mathematical convenience, which can be vaguely summarized by saying Φ admits the Dirac formalism. Two aspects of this formalism are: 1) All algebraic operations with the operators are allowed and no questions with respect to the domain of definition arise. 2) For every e.s.a. operator there exists a complete system of generalized eigenvectors. 1) follows from the fact that the elements of the algebra A are continuous operators with respect to τ_Φ and therefore uniquely defined on the whole space Φ[19].

To show this it is sufficient to prove that P and Q or a and a^+ are continuous because of the following theorem:<u>Theorem</u> : The product and sum of continuous operators are continuous operators.

Thus, in general, our algebra *A* is an algebra of continuous operators.

We recall that (in spaces in which the topology can be defined by the convergence of sequences (i.e. spaces with the first axiom of countability) an operator A is continuous if for all sequences $\{\varphi_\gamma\}$ with $\varphi_\gamma \to 0$ it follows that $A\varphi_\gamma \to 0$. To prove that a (and a^+) is τ_Φ continuous let us consider a and a^+ on Φ.

We use the <u>lemma</u> (proved in Appendix IV): For every $\varphi \in \Psi$

$$(\varphi, a(N+I)^p a^+\varphi) \leq \kappa(\varphi, (N+I)^{p+1}\varphi) \tag{1-41}$$

where κ is some constant , $\kappa < \infty$.

Proof: Let $\varphi_\gamma \to 0$, with respect to τ_Φ, as $\gamma \to \infty$ $\|\varphi_\gamma\|_p \to 0$ for every p

$$(\varphi_\gamma, (N+I)^p \varphi_\gamma) \to 0 \text{ for every p} \tag{1-42}$$

To show that $a^+\varphi_\gamma \to 0$, with respect to τ_Φ, we have to show that

$$\|a^+\varphi_\gamma\|_q \to 0 \tag{1–43}$$

for every q, i.e., that

$$(a^+\varphi_\gamma, (N+I)^q a^+\varphi_\gamma) = (\varphi_\gamma, a(N+I)^q a^+\varphi_\gamma) \to 0$$

for every q. By (1-41)

$$(\varphi_\gamma, a(N+I)^q a^+\varphi_\gamma) \leq \kappa(\varphi_\gamma, (N+I)^{q+1}\varphi_\gamma) \tag{1-44}$$

but by (1-42) the r.h.s. of this inequality $\to 0$, consequently also the l.h.s., which proves (1-43).

We remark that the convergence of $\|a^+\varphi_\gamma\|_q \to 0$ for $\gamma \to \infty$ for a fixed q follows from the convergence of $\|\varphi_\gamma\|_{q+1} \to 0$. Therefore in the case of a finite number of norms, a^+ will not be a continuous operator. As the topology in the case of a finite number of norms is equivalent to the topology given by the highest norm (i.e. sequences that converge with respect to one converge also with respect to the other and vice versa). This fact implies that a^+ cannot be a continuous operator in the Hilbert space topology, which is well known. The proof of the continuity of a is analogous. Therewith, we have shown that a and a^+ are continuous operators on the dense subspace Ψ in Φ and can therefore be uniquely extended to operators on the whole space Φ[20].

We denote the operators a,a^+, P, Q,...extended to the whole space Φ ,again by a,a^+, P, Q,..and domain questions do not arise in the algebraic operations.

The τ_Φ-continuous operators P, Q, H considered as operators on H are not closed operators [21] (they are τ_Φ–Continuous and consequently τ_Φ-closed but not τ_H -closed), because if $f_\gamma \in \Phi$ and $f_\gamma \to f$, with respect to τ_H , but $f \notin \Phi$, then $Af_\gamma \in \Phi$ for every γ but A is not defined on f. If $Af_\gamma \to g$, with respect to τ_H , then we define $\bar{A}f$ by $\bar{A}f = g$. We can do this for all $f \in H$ which are τ_H -limit points of some sequences $f_\gamma \in \Phi$ and for which Af_γ τ_H -converges. The operator $\bar{A}$ defined in this way, i.e. whose domain $D(\bar{A})$ is the set of all these $f \in H$, is the closure of A in the completion of Φ to H .

Thus, in correspondence to the relation (1-34) between the spaces: $H \supset \Phi$ we have the relation

$$\bar{A} \supset A \tag{1-45}$$

between the operators (however in general $D(\bar{A}) \neq H$)

So far we know of the operators P, Q, H (or N) that they are τ_Φ-continuous symmetric operators on Φ. Therefore,

$$P^+ \supset P \quad (\text{i.e., } D(P^+) \supset D(P)) \tag{1-46}$$

and $Q^+ \supset Q$

H and N are e.s.a. and therefore

$$H^+ = \bar{H} \quad \text{and} \quad N^+ = \bar{N} \tag{1-47}$$

However as a consequence of (1-47) it follows that also P and Q are e.s.a., i.e.

$$P^+ = \bar{P} \qquad D(P^+) = D(\bar{P}) \quad , \quad Q^+ = \bar{Q} \quad D(Q^+) = D(\bar{Q}) \tag{1-48}$$

<u>Remark</u> : H + I is, except for some constant factors , the Nelson operator, which is because of (1-47) e.s.a.. Therefore- as mentined before- P,Q, I are the group genrators of a group of transformations (Weyl group).

As a consequence of the fact that P, Q are elements of the Lie algebra of this group it follows then that P and Q are also e.s.a. by a theorem of Nelson and Stinespring[22)].

We now turn to the second aspect of the Dirac formalism, the existence of a complete set of genralized eigenvectors, which follows from the fact that for the operators on Φ the nuclear spectral theorem applies. The explanation of this requires some further mathematical preparations.

Definition : A bounded self-adjoint operator B is Hilbert-Schmidt if $B = \Sigma\lambda_k P_k$ where the projections P_k project on finite dimensional spaces H_k and $\Sigma\,(|\lambda_k| \dim H_k)^2 < \infty$.

Instead of giving the original definition of nuclear space[23)] we shall use a theorem of Roberts[23)] which gives a necessary and sufficient condition for Φ to be nuclear.

Definition : Φ is a nuclear space if there exists an e.s.a. τ_Φ-continuous operator $A \in \mathcal{A}$, whose inverse is Hilbert-Schmidt.

It is now very easy to see that our Φ is nuclear because N is e.s.a., the spectrum of $(N + I)^{-1}$ is $(n + 1)^{-1}$, n = 0, 1, 2,... and R_n is one dimensional, thus $\Sigma(n + 1)^2 < \infty$. Thus N + I is the operator that fulfills the above definition.

Remark concerning generalizations :

The construction of a nuclear space described above is immediately generalized to a more general algebra. The analogue of the lemma (1-41)

$$(\varphi,X(\Delta + I)^p X\varphi) \leq (\varphi,(\Delta + I)^{p+1}\varphi) \tag{1-49}$$

where X is one of the generators X_i and $\Delta = \Sigma X_i^2$ (Nelson operator) holds for all enveloping algebras (lemma by Nelson). Therwith the continuity of the algebra in a linear topological space in which the topology is defined by the countable number of scalar products

$$(\varphi,\psi)_p = (\varphi,(\Delta + I)^p\psi) \tag{1-50}$$

follows immediately. Further, if Δ is e.s.a. then all symmetric[12)] generators are also e.s.a. (theorems by Nelson and Nelson and Stinespring).

(1-49) is much stronger than required for the proof of the continuity of the generators. The continuity of the generators (and therewith the whole algebra) can already by proved if instead of p + 1 on the r.h.s. of (1-49) one has p + n, where n is any finite integer. Therefore, it appears that the continuity of the generators can already be proved for any finitely generated associative algebra.

The nuclearity is much harder property to establish. It has been proven for the cases that the algebra is the enveloping algebra E(G) of the following groups G: G nilpotent (because then E(G) is isomorphic to the enveloping algebra generated by P_α, Q_α , $\alpha = 1, 2, ...m$ with $[P_\alpha, Q_\beta] = -\delta_{\alpha\beta}I$ for some m (Theorem by Kirilov)[25] and we have just an m dimensional generalization of the above described case)

G semi-simple (A. Bohm) [26]

G = AΘK semidirect product, with A abelian and K compact. (B. Nagel)[27]

G = Poincaré group for some of the representations (B. Nagel)[27]

We can further generalize the notion of nuclearity to other spaces such as:

1- Subspaces of nuclear spaces.

2- Spaces which do not satisfy the first axiom of countability.

A general definition of nuclear space can be seen in References 4 and 24 at the end of this chapter. The nuclear spaces described so far are indeed particular instances of the general case. In references 4 and 24, we can also find important general results. One of the most important ones is that any subspace of a nuclear space is also nuclear. As a consequence, any closed subspace of a complete nuclear space is also a complete nuclear space. We shall make extensive use of this result in Chapter IV.

Examples of nuclear spaces which do not satisfy the first axiom of countability are Φ^x, to be defined in the next section, and the spaces K and Z to be defined in Chapter II.

III. THE CONJUGATE SPACE OF Φ

An antilinear functional F on a linear space Ψ is a function, the values $F(\varphi) = \langle \varphi | F \rangle$ of which are complex numbers, which satisfies:

$$F(\alpha\varphi + \beta\psi) = \alpha^* F(\varphi) + \beta^* F(\psi) \; ; \; \varphi, \psi \in \Phi, \alpha, \beta, \in \mathbb{C} \tag{1-51}$$

or, in the other notation

$$\langle \alpha\varphi + \beta\psi | F \rangle = \alpha^* \langle \varphi | F \rangle + \beta^* \langle \psi | F \rangle \tag{1-52}$$

The functional F is called τ_Φ–continuous if from

$\varphi_\gamma \to \varphi$, with respect to τ_Φ

follows

$$F(\varphi_\gamma) \to F(\varphi), \quad \text{with respect to C} \tag{1-53}$$

where $\to$, with respec to C, means convergence in the sense of complex numbers.

One can prove that F is τ_Φ-continuous if there exists a positive constant $\mathfrak{R}$ and a natural number q such that

$$|F(\varphi)| < \mathfrak{R} \, ||\varphi||_q \qquad \forall \, \varphi \in \Psi \tag{1-54}$$

Two antilinear functionals can be added so as to obtain another antilinear functional. Also, they can be multiplied by complex numbers. We can define these two operations by means of the following formula:

$$(\alpha F_1 + \beta F_2)(\varphi) = \alpha F_1(\varphi) + \beta F_2(\varphi) \tag{1-55}$$

or

$$\langle \varphi | \alpha F_1 + \beta F_2 \rangle = \alpha \langle \varphi | F_1 \rangle + \beta \langle \varphi | F_2 \rangle \tag{1-56}$$

The functional $\alpha F_1 + \beta F_2$, defined by (1-55) or (1-56), is an antilinear functional on Ψ. Thus, the set of antilinear functionals on an arbitrary linear space Ψ is again a linear space. This space is called the algebraic antidual or algebraic conjugate space of Ψ and is denoted by Ψ^x.

Continuous antilinear functionals have not to fulfill (1-51), but also (1-53). Therefore, the set of continuous antilinear functionals, which also forms a linear space, is smaller than the set Ψ^x and depends-like all topological notions- on the particular topology that has been chosen. We shall denote the set of all τ_Φ-continuous antilinear functionals by Φ^x.

On the other hand, let $f \in H$. f defines an antilinear functional on Ψ by

$$\langle \varphi | F \rangle = F(\varphi) := (\varphi, f). \tag{1-57}$$

It is easy to see that F defined by the vector f as given in (1-57) fulfills the condition (1-51) if (φ, f) fulfilles the conditions for a scalar product.

Assume now $\varphi_\gamma \to \varphi$, with respect to τ_Φ. This implies in particular that $\varphi_\gamma \to \varphi$ with respect to τ_H . Since in a Hilbert space every convergent sequence converges also weakly, then $(\varphi_\gamma, f) \to (\varphi, f)$. Therefore, $F(\varphi_\gamma) \to F(\varphi)$, if F is defined as in (1-57), and $F \in \Phi^x$. This is true, in particular , if $f \in \Psi$ (we are assuming that $\Psi \subset \Phi \subset H$. Recall section II of the present chapter)
Thus, the scalar product on Ψ (which is inhereted from that on H) defines a τ_Φ-continuous antilinear functional for every member of Ψ. Also, we have obtained that $H \subset \Phi^x$.

Then, we have the situation,

$$\Phi^{x} \supset H \supset \Phi \supset \Psi \tag{1–58}$$

We call the triplet

$$\Phi^{x} \supset H \supset \Phi \tag{1–59}$$

a Rigged Hilbert Space or Gel'fand triple (If Φ were a more general nuclear space we would be in the same situation. On the other hand, note that if F is antilinear τ_{Φ}–continuous functional on Ψ so is on its completion Φ and vice versa).

Whenever $f \in H$, we shall write $\langle \varphi | f \rangle = (\varphi, f)$. Thus, we identify the vector f with the functional F in our notation for the sake of simplicity.

We can endow Φ^{x} with various topologies and therewith various meanings of convergence of sequences. The most important example is the weak convergence.

<u>Definition</u> : A sequence of functionals $\{F_{\gamma}\}$ in Φ^{x} converges weakly or τ^{x}-converges to $F \in \Phi^{x}$ if

$$\langle \varphi | F_{\gamma} \rangle \to \langle \varphi | F \rangle, \text{ with respect to } C \tag{1-60}$$
$$\gamma \to \infty, \gamma \text{ natural}$$

for every $\varphi \in \Phi$.

We shall denote this sense of convergence by $F_{\gamma} \to F$, with respect to τ^{x}.

<u>Remark</u> : As a matter of fact, the so called weak topology [28),29)] on Φ^{x} cannot be completely described by the passage to the limit of countable sequences. In fact Φ^{x} does not satisfy the first axiom of countability and is, therefore, a linear topological space different from that one presented in the previous section.

One can show that Φ^{x} is τ^{x}-complete. We can define a τ^{x}-Cauchy sequence as follows : (F_{γ}) is a τ^{x}-Cauchy sequence if for any $\varphi \in \Phi$ and $\varepsilon > 0$, there exists a natural number N such that if $\mu,\nu >$ N, then $|\langle \varphi | (F_{\mu} - F_{\nu}) \rangle| < \varepsilon$. The τ^{x}-completeness of Φ^{x} implies, in particular, the existence of a τ^{x}-limit point in Φ^{x} for any τ^{x}-Cauchy sequence, the converse being not true.

Since we have not completely described the τ^{x} topology, we cannot attemp to prove the following statements:

Being Φ^{x} a linear topological space, we can consider τ^{x}-continuous antilinear functionals φ on Φ^{x} ,i.e., the set Φ^{xx} of the mappings from Φ^{x} into[30)] C which satisfy:

$$\varphi\,(\alpha F_1 + \beta F_2) = \alpha^{*}\ \widetilde{\varphi}\,(F_1) + \beta^{*}\ \widetilde{\varphi}\,(F_2) \tag{1-61}$$
$$F_1, F_2 \in \Phi^{x}, \alpha, \beta \in C$$

or, in the other notation:

$$\langle \alpha F_1 + \beta F_2 | \tilde{\varphi} \rangle = \alpha^* \langle F_1 | \tilde{\varphi} \rangle + \beta^* \langle F_2 | \tilde{\varphi} \rangle \tag{1-62}$$

and there exists a positive constant K and n fixed vectors in Φ, $\varphi_1, \varphi_2, ...\varphi_n$, such that

$$|\langle F | \tilde{\varphi} \rangle| \leq K (|\langle \varphi_1 | F \rangle| + |\langle \varphi_2 | F \rangle| + ... + |\langle \varphi_n | F \rangle|) \tag{1-63}$$

for any $F \in \Phi^X$. (1-63) implies the continuity of $\tilde{\varphi}$ on Φ^X.[31] As a consequence if

$F_\gamma \to F$, with respect to τ^X

then

$$\tilde{\varphi}(F_\gamma) \to \tilde{\varphi}(F), \text{ with respect to C} \tag{1-64}$$

Φ^{XX} is a linear topological space if addition of functionals and multiplication by scalars are defined by:

$$\langle F | \alpha \tilde{\varphi}_1 + \beta \tilde{\varphi}_2 \rangle = \alpha \langle F | \tilde{\varphi}_1 \rangle + \beta \langle F | \tilde{\varphi}_2 \rangle \tag{1–65}$$

and weak convergence of sequences by

$$\tilde{\varphi}_\gamma \to \tilde{\varphi} \quad \text{if } \langle F | \tilde{\varphi}_\gamma \rangle \to \langle F | \tilde{\varphi} \rangle \tag{1-66}$$

for every $F \in \Phi^X$.

In addition, one can prove that to each element $\varphi \in \Phi$ corresponds one and only one continuous antilinear functional on Φ^X defined by the equation

$$\tilde{\varphi}(F) = F(\varphi)^* \quad \text{or} \quad \langle F | \tilde{\varphi} \rangle = \langle \varphi | F \rangle^* \tag{1-67}$$

the converse being also true. The mapping $\varphi \to \tilde{\varphi}$ given by (1-67) is, therefore, one to one and onto. It is also linear. That means that the linear spaces Φ and Φ^{XX} are, rougly speaking, the same one. Moreover, the weak topology on Φ^{XX} coincides with τ_Φ. Therefore, both linear topological spaces coincide and we can write:

$$\Phi = \Phi^{XX} \tag{1-68}$$

We say that Φ is reflexive because of (1-68).

Another important property is the following: Φ, and therefore H is τ^X-dense in Φ^X. This means that the τ^X-closure of Φ is Φ^X. Equivalently, if $F \in \Phi^X$ any τ^X-neighborhood of F in Φ^X contains points of Φ.

<u>Definition</u> : A continuous operator B^X on Φ^X is a mapping from Φ^X into itself such that

$$B^X(\alpha F_1 + \beta F_2) = \alpha B^X(F_1) + \beta B^X(F_2) \tag{1-69}$$

$F_1, F_2 \in \Phi^x$; $\alpha, \beta \in C$

and for every $\varphi \in \Phi$ there are a positive constant K and n fixed vectors in Φ, $\varphi_1, \varphi_2, ..., \varphi_n$, such that:

$$|\langle\varphi|B^xF\rangle| \leq K\,(|\langle\varphi_1|B^xF\rangle| + ... + |\langle\varphi_n|B^xF\rangle|) \qquad (1\text{–}70)$$

(1–69) shows that B^x is a linear mapping, (1-70) that it is continuous. [31] As a consequence of the continuity of B^x, if $F_\gamma \to F$, with respect to τ^x, then $B^xF_\gamma \to B^xF$, with respect to τ^x. Since Φ^x does not satisfy the first axiom of countability, the converse is not true.

Let A be a τ_Φ-continuous e.s.a. operator on Φ. We can define its extension A^x into Φ^x as

$$\langle\varphi|A^xF\rangle = F(A\varphi) = \langle A\varphi|F\rangle \qquad (1\text{–}71)$$

As here defined, A^x is a continuous operator on Φ^x. In particular the extensions into Φ^x of P, Q, and H, which respectively are P^x, Q^x and H^x, are τ^x-continuous operators on Φ^x. Moreover, for every τ_Φ-continuous e.s.a. operator A, we have, in correspondence to the relation between the spaces

$$\Phi^x \supset \mathcal{H} \supset \Phi^x \qquad (1\text{-}72)$$

the relation between the operators

$$A^x \supset \overline{A} \supset A \qquad (1\text{-}73)$$

where $\overline{A}$ is the closure of A in $\mathcal{H}$ (see Appendix III).

If A is not e.s.a. on Φ, but densely defined on $\mathcal{H}$, its Hilbert space adjoint A^+ is still well defined on $\mathcal{H}$. If further, A^+ is defined on every vector of Φ, the corresponding images are in Φ (i.e., $\Phi \supset A^+\Phi$ and then we shall say that Φ is stable under the action of A^+) and A^+ is τ_Φ-continuous, we still can define the extension of A, A^x, on Φ^x by

$$\langle\varphi|A^xF\rangle = \langle A^+\varphi|F\rangle\ ;\ \forall\,\varphi \in \Phi\ ;\ F \in \Phi^x \qquad (1\text{-}74)$$

We can prove that A^x is τ^x-continuous.

If A is e.s.a. on Φ (or even symmetric) $A\varphi = A^+\varphi \in \Phi$. Therefore, (1-71) can be looked as a particular case of (1-74).

IV. GENERALIZED EIGENVECTORS AND THE NUCLEAR SPECTRAL THEOREM

Before we define the notion of generalized eigenvectors of an operator A on the space Φ, let us recall the situation in the Hilbert space $\mathcal{H}$.

A nonzero vector $h \in \mathcal{H}$ is called an eigenvector of an operator A on $\mathcal{H}$ if

$$Ah = ah \tag{1-75}$$

where a is a complex number. In general, it is not possible to find for every operator on $\mathcal{H}$ an eigenvector. In fact, it is well known that the operators P and Q do not have any eigenvector in $\mathcal{H}$. Yet physicists, following Dirac, always work with "eigenvectors" of P and Q

$$P|p\rangle = p|p\rangle\,;\ Q|q\rangle = q|q\rangle \tag{1-76}$$

and use the assumption that these eigenvectors form a "complete" system meaning that

$$\int |p\rangle\langle p|\,dp = 1 \tag{1-77}$$

in some sense. Here, we shall give a definition and a theorem that provides a mathematical justification for this procedure.

<u>Definition</u> : Let A be an operator on Φ, which can be extended to $\Phi^{\times}$. In particular A can be an operator satisfying the requirements written at the end of the previous section. A generalized eigenvector of the operator A corresponding to the generalized eigenvalue λ is an antilinear functional, $F \in \Phi^{\times}$, such that

$$F(A^{+}\varphi) = \langle A^{+}\varphi|F\rangle = \langle\varphi|A^{\times}F\rangle = \lambda\langle\varphi|F\rangle \tag{1–78}$$

holds for every $\varphi \in \Phi$. This may be written in the form

$$A^{\times}F = \lambda F \tag{1-79}$$

or following Dirac

$$A^{\times}|F\rangle = \lambda|F\rangle \tag{1-80}$$

Let us assume A has a generalized eigenvector in the Hilbert space, i.e., F in (1-78) is an element of $\mathcal{H}$. Then (1-78) reads

$$\langle A^{\times}\varphi|F\rangle = (A^{+}\varphi, f) = (\varphi, Af) = \lambda(\varphi, f) \tag{1-81}$$

for every $\varphi \in \Phi$ and consequently (since Φ is $\tau_{\mathcal{H}}$ -dense in $\mathcal{H}$)

$$Af = \lambda f. \tag{1-82}$$

Thus, a generalized eigenvector which is an element of the Hilbert space is an ordinary eigenvector corresponding to the same eigenvalue.

Before stating the nuclear spectral theorem, let us present a few remarks:

Every self-adjoint operator A on a finite dimensional Hilbert space has a complete system of eigenvectors $h_i = |\lambda_i) \in \mathcal{H}$ such that

$$Ah_i = \lambda_i h_i\ ;\ (h_i, h_j) = (\lambda_i|\lambda_j) = \delta_{ij} \tag{1-83}$$

and for every $h \in \mathcal{H}$

$$h = \sum_{i=1}^{n} |\lambda_i)\,(\lambda_i\,|h) \quad n = \dim H$$

(1-84)

The set $\Lambda = \{\lambda_1, \lambda_2,...,\lambda_n\}$ of the eigenvalues of A is called the spectrum of A.

For an infinite dimensional Hilbert space this statement is not longer true.

In fact, we know that they are operators that have no eigenvectors in H (e.g. Q and P). The spectrum of an operator A on H is the set Λ of complex numbers which do not belong to the resolvent set $\rho(A)$. Then we define $\rho(A)$ as the set of complex numbers μ such that:

1. $(A - \mu I)$ has an inverse on H, i.e., it is one-to-one.
2. Its inverse $(A - \mu I)^{-1}$ is densely defined on H, i.e., $(A - \mu I)H$ is dense in H.
3. $(A - \mu I)^{-1}$ is τ_H -continuous on H.

If f is an eigenvalue of A in H, $Af = \mu f$ for some complex number μ. Thus $(A - \mu I)f = 0$ and $A - \mu I$ has a nonzero kernel. A well known result in elementary algebra says that $A - \mu I$ cannot be one-to one and, therefore, μ cannot belong to $\rho(A)$. Consequently, $\mu \in \Lambda$, the spectrum of A.

We have shown that any eigenvalue of A belongs to the spectrum of A. However, not every element in the spectrum is an eigenvalue. P and Q have an spectrum which coincide with the real line in both cases. Nevertheless, we know the set of their eigenvalues in H is empty.

The spectrum of a self-adjoint operator A is a closed subset of the real axis. If the self adjoint operator A on the infinite dimensional Hilbert space H has only discrete spectrum, i.e., the spectrum coincides with the set of its eigenvalues $\Lambda = \{\lambda_1, \lambda_2,...,\lambda_n,....\}$, then the above statement for operators on finite dimensional Hilbert spaces carries over to the infinite dimensional case. To avoid complications which are inessential for the principal problems and inapplicable for the special problem of the one-dimensional harmonic oscillator, we restrict ourselves to cyclic operators.

Definition : An operator is cyclic if there exists a vector $f \in H$ such that $A^k f$, k = 0, 1, 2,....... span the entire Hilbert space. It is clear that the operators P and Q are cyclic because

$$Q^k \varphi_0 = \left(\frac{\hbar}{2m\omega}\right)^{k/2} (a + a^+)^k \varphi_0$$

(1-85)

and

$$P^k \varphi_0 = 2i\left(\frac{1}{m\hbar\omega}\right)^{k/2} (a - a^+)^k \varphi_0$$

$k = 0,1,2,....$

span H .

The generalization of the above statement to the case of a cyclic self-adjoint operator with discrete spectrum is given by the following spectral theorem:

There exists an orthonormal set of eigenvectors

$$h_i = |\lambda_i) \in H$$
$$Ah_i = \lambda_i h_i \ ; \ \lambda_i \in \Lambda \qquad (1\text{-}86)$$
$$(h_i, h_j) = (\lambda_i|\lambda_j) = \delta_{ij}$$

such that for every $h \in H$

$$h = \sum_{i=1}^{n} h_i (h_i, h) = \sum_{i=1}^{\infty} |\lambda_i)(\lambda_i|h)$$

$$\int_\Lambda d\mu(\lambda)\ |\lambda)(\lambda|h)$$

(1-87)

The measure [32] μ is concentrated on the eigenvalues $\lambda_i \in \Lambda$, i.e., $\mu(\Lambda) = 1, \mu\{C - \Lambda\} = 0$.

The Hamiltonian of the one dimensional harmonic oscillator H is an example of such an operator. We remark that all the eigenvectors of A appear in the above decomposition.

In the general case, there is not possible such a decomposition. However, a generalization of the above statement is the subject of the Nuclear Spectral Theorem, also called the Gelfand Maurin theorem.

The Nuclear Spectral Theorem (Gelfand-Maurin)

Let $\Phi \subset H \subset \Phi^x$ be a rigged Hilbert space and A a cyclic e.s.a. τ_Φ-continuous operator. Then, there exists a set of generalized eigenvectors

$$F_\lambda = |\lambda\rangle$$
$$A^x |\lambda\rangle = \lambda |\lambda\rangle; \ \lambda \in \Lambda, \quad \text{i.e.,}$$

$$\langle A\varphi|F_\lambda\rangle = \langle\psi|A^x F_\lambda\rangle = \lambda\langle\varphi|F_\lambda\rangle. \ \forall \in \Phi \qquad (1\text{-}88)$$

such that for every $\varphi, \psi \in \Phi$ and some uniquely defined positive measure μ on Λ

$$(\psi, \varphi) = \int_\Lambda d\mu(\lambda)\ \langle\psi|F_\lambda\rangle\langle F_\lambda|\varphi\rangle = \int_\Lambda d\mu(\lambda)\langle\psi|\lambda\rangle\langle\lambda|\varphi\rangle$$

with (1–89)

$$\langle F_\lambda|\varphi\rangle = \langle\varphi|F_\lambda\rangle^* = \langle\lambda|\varphi\rangle; \quad \varphi \in \Phi \qquad (1\text{-}90)$$

(1-88) is written formally

$$\varphi = \int_{\Lambda} d\mu(\lambda)\, |\lambda\rangle\langle\lambda|\varphi\rangle$$

(1-91)

Furthermore, if f(λ) is a well behaved function on Λ:

$$(\psi, f(A)\varphi) = \int_{\Lambda} d\mu(\lambda)\langle\psi|\lambda\rangle\langle\lambda|\varphi\rangle f(\lambda)$$

(1-92)

which is written formally

$$f(A) = \int_{\Lambda} d\mu(\lambda)\, |\lambda\rangle\langle\lambda| f(\lambda)$$

(1-93)

If f(λ) = 1, f(A) is obviously the identity operator I because of (1-89). If f(λ) = λ, f(A) = A and (1-93) yields in these two cases

$$I = \int_{\Lambda} |\lambda\rangle\langle\lambda| d\mu(\lambda)$$

(1-94)

$$A = \int_{\Lambda} \lambda\, |\lambda\rangle\langle\lambda| d\mu(\lambda)$$

(1-95)

Note that (1-77) is a particular case of (1-93) with $\Lambda = R$; $\lambda = p \in R$ and $d\mu(\lambda) = d\lambda$ the Lebesgue measure on R.

If we set $\psi = \varphi$ in (1-89), then we see that from the vanishing of all components $\langle\lambda|\varphi\rangle$ of the spectral decomposition of φ corresponding to the operator A, it follows that $\| \varphi \| = 0$, i.e., $\varphi = 0$. Because of this property, the set of generalized eigenvectors $|\lambda\rangle$ ocurring in (1-89) and (1-92) is called complete in analogy to the completeness of the system of ordinary eigenvectors in (1-87).

The spectral theorem assures the existence of a complete set of generalized eigenvectors of the operators Q and P, i.e., of generalized vectors that fulfill

$$Q^{x}|x\rangle = x|x\rangle \,; \quad P^{x}|p\rangle = p|p\rangle \qquad (1\text{-}96)$$

$$\varphi = \int_{\text{spectrum } Q=R} d\mu(x)\; |x\rangle\langle x|\varphi\rangle$$

(1-97)

$$= \int_{-\infty}^{\infty} dx\, |x\rangle\langle x|\varphi\rangle$$

or

$$\varphi = \int_{\text{spectrum } P=R} d\mu(p)\, |p\rangle\langle p|\varphi\rangle$$

(1-98)

$$= \int_{-\infty}^{\infty} dp\, |p\rangle\langle p|\varphi\rangle$$

The fact that there exists a complete set of eigenvectors does not tell us what this spectrum is. As we have mentioned already, the spectrum of P and Q is the real line in both cases. However, it is not so widely known that the derivation of this is far from trivial.[33)] The reason for this is that the problem of the physicists was the reverse of the prblem described here, namely to find the defining assumptions (1-1) and (1-2) from the spectrum of P and Q which were assumed to be the real line.

In general, an e.s.a. operator has more generalized eigenvectors than appear in the spectral decomposition. The spectral decomposition itself does not need to be unique. In particular the set of generalized eigenvalues of P and Q may be the complex plane.

The Gelfand-Maurin theorem is also valid in the general case, where Φ is a general nuclear space. Even if we keep Φ as defined so far, this result is important and deserves further comments.

First of all, $\langle\lambda|\varphi\rangle$ is a complex function of the real variable $\lambda \in \Lambda$. If we write

$$(\varphi, \varphi) = \int_{\Lambda} d\mu(\lambda)\langle\varphi|\lambda\rangle\langle\lambda|\varphi\rangle = \int_{\Lambda} d\mu(\lambda)\, \overline{\langle\lambda|\varphi\rangle}\, \langle\lambda|\varphi\rangle = \int_{\Lambda} d\mu(\lambda)\, |\langle\lambda|\varphi\rangle|^2$$

(1-99)

we immediately realize that $\langle\lambda|\varphi\rangle = \varphi(\lambda)$ is a square integrable function with respect to $d\mu(\lambda)$. We shall denote this space as $L^2(\Lambda, d\mu(\lambda))$[32)].

On the other hand, observe that the scalar product on $L^2(\Lambda, d\mu(\lambda))$ is

$$(\varphi(\lambda), \psi(\lambda))_{\Lambda} = \int_{\Lambda} d\mu(\lambda)\, \overline{\varphi(\lambda)}\, \psi(\lambda)$$

(1–100)

and if $\varphi, \psi \in \Phi$, we have

$$(1\text{-}100) = \int_{\Lambda} d\mu(\lambda)\langle\varphi|\lambda\rangle\langle\lambda|\varphi\rangle = (\varphi, \psi)$$

(1–101)

and therefore, the correspondence $\varphi \leftrightarrow \varphi(\lambda)$ preserves the scalar product on Φ. This correspondence is obviuosly linear because $\langle\lambda|\varphi+\psi\rangle = \langle\lambda|\varphi\rangle + \langle\lambda|\psi\rangle$ and for $\omega \in C$, $\langle\lambda|\omega\varphi\rangle = \omega\langle\lambda|\varphi\rangle$. Thus, it can be represented by means of an operator U such that $U\varphi = \varphi(\lambda)$; $\forall\varphi \in \Phi$. Since U preserves scalar products (see (1-101)), it is an isometry. Moreover, one can prove that the image of Φ by U is dense in $L^2(\Lambda, d\mu(\lambda))$. Therefore, its unique extension to U is a unitary operator onto $L^2(\Lambda, d\mu(\lambda))$. If $f \in H$, $Uf = f(\lambda) \in L^2(\Lambda, d\mu(\lambda))$, but it is important to remark that $f(\lambda) = \langle\lambda|f\rangle$ if and only if $f \in \Phi$, otherwise $\langle\lambda|f\rangle$ makes no sense.

Hereafter, we call $U\Phi$ the image of Φ by U. We can endow $U\Phi$ with a topology τ_λ. If Φ satisfies the first axiom of countability,[34)] which we shall assume from now on, so does $U\Phi$ and its topology can be defined as follows:

$$f_n(\lambda) \to f(\lambda), \text{ with respect to } \tau_\lambda \text{ iff } f_n \to f, \text{ with respect to } \tau_\Phi \tag{1-102}$$

where $Uf_n = f_n(\lambda)$ and $Uf = f(\lambda)$. All topological properties are transferred from Φ into $U\Phi$ by means of U. In particular, $U\Phi$ is a τ_λ-complete nuclear space. Furthermore, it is dense in $L^2(\Lambda,d\mu(\lambda))$ in the sense of convergence in the latter, i.e.,for any $f(\lambda) \in L^2(\Lambda,d\mu(\lambda))$ there exists a sequence of functions in $U\Phi$, $\{f_n\}$, such that

$$\int_\Lambda d\mu(\lambda)\ |f_n(\lambda) - f(\lambda)|^2 \to 0 \tag{1-103}$$

Our next step will be to extend U to $\Phi^\times$. We call $U^\times$ to this extension. In order to do it, we use the prescription

$$\langle U\varphi|U^\times F_\lambda\rangle = \langle\varphi|\lambda\rangle \tag{1-104}$$

Where here $F_\lambda = |\lambda\rangle$. $U^\times F_\lambda$ is indeed an antilinear functional on $U\Phi$. We leave the proof to the reader. On the other hand, $U^\times$ really extends U. This means that if $f \in H$ then $U^\times f = Uf$. In fact,

$$\langle U\varphi|U^\times f\rangle = \langle\varphi|f\rangle = (\varphi,f) = (U\varphi,Uf) = \langle U\varphi|Uf\rangle \tag{1–105}$$

for any $\varphi \in \Phi$, or equivalently for any $U\varphi \in U\Phi$. Therefore, the functionals $U^\times f$ and Uf are just the same.

Furthermore, $U^\times F_\lambda$ is a continuous antilinear functional on $U\Phi$, for if $U\varphi_n \to U\varphi$

$$\langle U\varphi_n|U^\times F_\lambda\rangle = \langle\varphi_n|\lambda\rangle \to \langle\varphi|\lambda\rangle = \langle U\varphi|U^\times F_\lambda\rangle \tag{1–106}$$

Therefore, $U^\times F_\lambda \in (U\Phi)^\times$ this space being the antidual of $U\Phi$. One can also prove that $U^\times : \Phi^\times \to \Phi^\times$ is continuous. Hence

$$U^\times\Phi^\times = (U\Phi)^\times \tag{1-107}$$

and both spaces have the same toplogical structure.

Summarizing, starting with $\Phi \subset H \subset \Phi^\times$ and using the unitary operator U, we have constructed a new rigged Hilbert space

$$U\Phi \subset L^2(\Lambda,d\mu(\lambda)) \subset U^\times\Phi^\times \tag{1-108}$$

<u>Definition</u> : Two rigged Hilbert spaces $\Phi \subset H \subset \Phi^\times$ and $\Psi \subset G \subset \Psi^\times$ are equivalent if and only if there exists a unitary operator U on H such that

1. $UH = G$

2. $U\Phi = \Psi$ and U is bicontinuous with respect to the topologies on Φ and Ψ.

As a consequence U can be extended to Φ^{x} as in (1-104). Its extension U^{x} is a bicontinuous bijective linear mapping from Φ^{x} onto Ψ^{x} and hence $U^{x}\Phi^{x} = \Psi^{x}$.

If $\Phi \subset H \subset \Phi^{x}$ is an abstract rigged Hilbert space and $\Psi \subset G \subset \Psi^{x}$ is an equivalent one in which G is a Hilbert space of functions, the second r.H.s. is called a representation of the former. In particular, (1-108) is a representation of our original triplet. The vectors on Ψ^{x} are called generalized functions.

As we have mentioned earlier, the Gelfand-Maurin theorem for Q yields $\Lambda = R$ and $d\mu(\lambda) = d\lambda$.[35] Thus, in this case $L^2(\Lambda, d\mu(\lambda)) = L^2(R)$ and U is a unitary operator from H onto $L^2(R)$. $U\Phi$ is here S, the linear space of all real continuous functions f(x) on R having continuous derivatives to all orders and such that

$$\lim x^n \frac{d^m}{dx^m} f(x) \rightarrow 0\,;\; x \rightarrow \pm\infty$$

$$n, m = 0,1,2,... \tag{1-109}$$

i.e., f(x) and all its derivatives vanish at $\pm\infty$ faster than the inverse of any polynomial on x.

S is a countably normed space. A sequence $\{f_n\} \subset S$ converges to $f(x) \in S$ if one of the following equivalent prescriptions holds

$$1.\quad \left\|\left(x^2 + \frac{d^2}{dx^2}\right)^m (f_n(x) - f(x))\right\| \rightarrow 0$$

$$m = 0,1,2,... \tag{1-110}$$

$$2.\quad \left\|\left(ax^2 + b\frac{d^2}{dx^2} + c\right)^m (f_n(x) - f(x))\right\| \rightarrow 0$$

$$m = 0,1,2,... \tag{1-111}$$

where a, b , c are constants.

$$3.\quad \left\|x^m \frac{d^k}{dx^k} (f_n(x) - f(x))\right\| \rightarrow 0$$

$$m,k = 0,1,2,... \tag{1-112}$$

where $\|g(x)\|^2 = \int_\Lambda |g(x)|^2\, dx$, i.e., the usual norm on $L^2(R)$

$$4.\quad \text{supreme}\, \left|x^m \frac{d^k}{dx^k} (f_n(x) - f(x))\right| \rightarrow 0$$

m, k = 0,1,2,... (1-113)

On the other hand, $U^{\times}\Phi^{\times} \subset S^{\times}$ is the space of the tempered distributions. [36] Thus, we have a new r.H.s.

$$S \subset L^2(R) \subset S^{\times} \qquad (1\text{-}114)$$

which is a realization of the triplet defined in Section III. The image by U of the function φ_n as defined in (1-13) is the n-th normalized Hermite function.

Along to the realization of the vectors of the triplet $\Phi \subset H \subset \Phi^{\times}$, we can speak about a realization of observables. This is performed by using operators on (1-114), or more generally on $U\Phi \subset L^2(\Lambda,d\mu(\lambda)) \subset U^{\times}\Phi^{\times}$. If A is an operator on Φ, UAU^{-1} is the corresponding operator on $U\Phi$. We call UAU^{-1} a realization of A.

In particular, if we choose the realization (1-114) for vectors, for any $f(\lambda) \in S$, we have

$$(UQU^{-1})\, f(\lambda) = \lambda f(\lambda) \qquad (1\text{-}115)$$

i.e., this realization of the position operator is the operator on S consisting in multiplying by λ every function in S. Also,

$$(UPU^{-1})\, f(\lambda) = -i\, d/d\lambda\ (f(\lambda)) \qquad (1\text{-}116)$$

The elements of the algebra *A* , at the beginning of Section I, can therefore, be realized by polynomials in λ and $-i\, d/d\lambda$.

Now, let us recall formula (1-91). It means that, at least formally, any vector in Φ can be written in terms of the generalized eigenvectors of A. Therefore. the eigenvectors of $A^{\times}$ form a system of generators for the space Φ, under some point of view. This system of generators is not a basis for Φ in the usual sense since : 1) the vectors $|\lambda\rangle$ do not belong to Φ in general; 2) the finite or countable infinite sum is replaced by an integral in which the wave function, $\langle\lambda|\varphi\rangle = \varphi(\lambda) = U\varphi$ plays the role of set of coefficients. Under this consideration, $\langle\lambda|\varphi\rangle$ is the coeficient for $|\lambda\rangle$ in the span (1-91) of φ. .We here recall which the situation in the Hilbert space *H* is. For any $f \in H$, we can write $f = \Sigma\, a_n f_n$, where (f_n) is an orthonormal basis in *H* . Thus, any $f \in H$ can be written in terms of a sequence of complex numbers (a_n) having the property $\Sigma\, |a_n|^2 < \infty$. Then, the Hilbert space *H* can be represented by the Hilbert space of the complex sequences with the mentioned property[37]. We represent Φ by the space of functions $U\Phi$, $\langle\lambda|\varphi\rangle$ for a fixed $\lambda \in \Lambda$ can be viewed as the "λ–th" component of φ in this system of generators.

At this point, it is perhaps important to stress that the unitary operator U depends drastically on A. First of all, the dependence of the space $L^2(\Lambda,d\mu(\lambda))$ on A is double. Both Λ and $d\mu(\lambda)$ depend on A. On the other hand, it can also happen that two different observables A and B give the same space $L^2(\Lambda,d\mu(\lambda))$, U being different in both cases. For instance, let A = Q and B = P. In both cases $L^2(\Lambda,d\mu(\lambda)) = L^2(R)$. However, the Gelfand-Maurin theorm for Q gives an unitary operator U from *H* onto $L^2(R)$, and for P the operator $V = F\, U$, where *F* is the operator Fourier transform to be defined in Chapter II, Section III.

Finally, let $\varphi \in \Phi$ and $F \in \Phi^{\times}$, $U\Phi = S$.

We write

$$\langle F|\varphi\rangle = \langle U^{x}F|U\varphi\rangle = \langle U^{x}F|\varphi(\lambda)\rangle \tag{1-117}$$

which is often written in the integral notation as[38]:

$$\int_{\Lambda} (U^{x}F)\,\varphi(\lambda)\,d\mu(\lambda) = \int_{\Lambda} (U^{x}F)\,\langle\lambda|\varphi\rangle d\mu(\lambda) \tag{1-118}$$

This is a symbol to denote $\langle F|\varphi\rangle$. The following notation is also used:

$$\langle F|\varphi\rangle = \int_{\Lambda} \langle F|\lambda\rangle\,\langle\lambda|\varphi\rangle\,d\mu(\lambda) \tag{1-119}$$

Thus $(U^{x}F)^{*} = \langle F|\lambda\rangle$ or $U^{x}F = \langle\lambda|F\rangle = \langle F|\lambda\rangle^{*}$. $\langle\lambda|F\rangle$ is no more than a symbol for $U^{x}F \in U^{x}\Phi^{x}$ stressing on its dependence on λ and hence on its generalized function character,

For instance, if $A = Q$

$$\langle F|\varphi\rangle = \int_{\Lambda} \langle F|x\rangle\,\langle x|\varphi\rangle\,dx \tag{1-120}$$

where $|x\rangle$ is the generalized eigenvector of Q with eigenvalue $x \in R$ ($Q^{x}|x\rangle = x\,|x\rangle$).

If $F = |x'\rangle$

$$\langle x'|\varphi\rangle = \int_{\Lambda} \langle x'|x\rangle\,\langle x|\varphi\rangle\,dx \tag{1-121}$$

Since $\langle x'|\varphi\rangle$ is the value of the function $U\varphi = \varphi(x) \in S$ at the point $x' \in R$, $\langle x'|x\rangle$ represents the Dirac delta generalized function $\delta(x - x')$.

V. A REMARK CONCERNING GENERALIZATION

For the case of a more general algebra of linear operators the e.s.a. operators are not cyclic (more than one quantum number is needed to characterize pure states). One then needs a complete system of commuting operators to obtain a complete set of generalized eigenvectors.

$\{A_k\}$, $k = 1,2,\ldots,N$ is a system of commuting operators iff

1 $[A_i, A_k] = 0$ for all $i,k = 1,\ldots,N$ (1-122-1)

2 $\Sigma\ A_k{}^2$ is e.s.a. (1-122-2)

$\{A_k\}$ is a complete commuting system if there exists a vector $\varphi \in \Phi$ such that

$\{A\varphi \mid A$ runs out the algebra generated by $\{A_k\}\}$

spans H (1-123)

An antilinear functional F on Φ is a generalized eigenvector for the system A_k if for any k = 1,...,N

$$(A_k)^{\times}F = \lambda^{(k)}F \qquad (1\text{-}124)$$

The set of numbers $\lambda = (\lambda^{(1)}, \lambda^{(2)}, \ldots, \lambda^{(N)})$ are called generalized eigenvalues corresponding to the generalized eigenvector $F_\lambda = |\lambda^{(1)}, \lambda^{(2)},\ldots, \lambda^{(N)}\rangle$. There exits a complete system of generalized eigenvectors according to the general form of the:

Nuclear Spectral Theorem

Let $\{A_k\}$, $k = 1,2,\ldots,N$ be a complete system of commuting e.s.a. τ_Φ-continuous operators on the rigged Hilbert space $\Phi \subset H \subset \Phi^{\times}$.

Then, there exists a set of generalized eigenvectors

$$|\lambda^{(1)}, \lambda^{(2)},\ldots,\lambda^{(N)}\rangle \in \Phi^{\times}$$

$$(A_k)^{\times}\, |\lambda^{(1)}, \lambda^{(2)},\ldots, \lambda^{(N)}\rangle = \lambda^{(k)}\ |\lambda^{(1)}, \lambda^{(2)},\ldots, \lambda^{(N)}\rangle \qquad (1\text{-}125)$$

$$\lambda^{(k)} \in \Lambda^{(k)} = \text{spectrum of } A_k$$

such that for every $\varphi \in \Phi$ and some uniquely defined measure μ on $\Lambda = \Lambda^{(1)} \times \Lambda^{(2)} \times \ldots \times \Lambda^{(N)}$, the Cartesian product of the $\Lambda^{(k)}$

$$(\psi|\varphi) = \int_\Lambda d\mu(\lambda)\, \langle\psi|\lambda^{(1)}, \lambda^{(2)},\ldots, \lambda^{(N)}\rangle\, \langle\lambda^{(1)}, \lambda^{(2)},\ldots, \lambda^{(N)}|\varphi\rangle$$

$\forall\ \psi, \varphi \in \Phi$ (1–126)

or written in another form,

$$\varphi = \int_\Lambda d\mu(\lambda)\ |\lambda^{(1)}, \lambda^{(2)},\ldots, \lambda^{(N)}\rangle\ \langle\lambda^{(1)}, \lambda^{(2)},\ldots, \lambda^{(N)}\ |\varphi\rangle$$

(1-127)

This theorm gives the precise formulation of the famous Dirac conjecture[2)] if the starting point is a precisely defined algebra.

The mathematical task that has to be solved if one starts with a well defined algebra is to find a complete commuting system and its spectrum. The problem to determine when a system is complete is far from trivial. Already for the symplest cases of enveloping algebras of group representations the number N is not independent of the particular commuting system, i.e., if $A_1,...,A_N$ is one particular commuting system, then another commuting system containing an operator B_1 may require N + 1 operators $B_1, B_2, ..., B_{N+1}$ in order to be complete.[39)]

The problem of a physicist is usually the reverse of this mathematical problem. From the experimental data he finds out how many quantum numbers are required, and what are the possible values of these quantum numbers. This gives him the complete commuting system $\{A_k\}$, a minimun number of other operators such the matrix elements of A calculated from the properties of this algebra agree with the experimental values of the corresponding observables.

Summarizing : Using the rigged Hilbert space we have reproduced the main features of the Dirac formalism. For the particular case of the harmonic oscillator we have in fact reproduced everything of Dirac's formalism. The difference between this formulation and the usual Hilbert space formulation appears to be minor from a physicist point of view, but it is essential from a mathematical point of view and leads to tremendous mathematical simplification; in fact, it justifies the mathematically undefined operations that physicists have been accustomed to in their calculations.

References on Chapter I

1. J. von Neumann, Mathematische Grundlagen der Quantenmechanik; Springer, Berlin, 1932; English version: "Methematical Foundations of Quantum Mechanics" , Princeton University Press, 1955.
2. P.A.M. Dirac,"The principles of Quantum Mechanics", Clarendon Press, Oxford 1958.
3. I.M. Gel'fand, G.F. Shilov,"Generalized Functions", vol II, in particular Chapter I, Academic Press, New York 1967.
   I.M. Gelfand, N.J.Vilenkin, "Generalized Functions", vol IV, in particular Chapter I.
4. K. Maurin, "General Eigenfunction Expansions and Unitary Representations of Topological Groups", Polish Scientific Publishers, Warsaw 1968.
5. J.E. Roberts, J. Math. Phys. 7, 1097 (1966). J.P. Antoine, J. Math. Phys. 10 ,53 (1969).
6. A. Bohm, Boulder Lectures in Theoretical Physics, vol. 9A, 255 (1965).
7. P. Kristensen, L. Melibo, E. Thue Poulsen, Commun. Math. Phys. 1, 175 (1965).
8. The explanation of the content of this basic assumption is the subject of this chapter. All the unknown mathematical notions will be defined or described in the following sections: Appendices I (Linear Space), II (Algebra), III (Operators in Hilbert Space).
9. The formulation given in this chapter is a special case of the construction of rigged Hilbert spaces for enveloping algebras of Lie groups which has been suggested in A. Bohm, J. Math. Phys. 8, 1557 (1966), Appendix; and independently by B. Nagel, College de France Lectures (1970). A detailed treatement of ε(SU(1,1)), which is slightly more complicated than the example treated here, has been given in G. Lindblad, B. Nagel, Ann. Institut Poincaré, Section A - (N.S.) 13, 27 (1970).
10. See Appendix II.
11. See Appendix I.
12. See Appendix III.
13. E. Nelson, Ann. Math. 70, 572 (1959).
14. H.D. Doebner, Proceedings of the 1966 Istanbul Summer Institute.
15. See Reference 13.
16. In the scalar product space, the norm $\|\ \|$ is defined by $\|\varphi\|^2 = (\varphi,\varphi)$.
17. The construction of the completion $\hat{R}$ of the space R consitst of the following : The elements of the space $\hat{R}$ are all possible Cauchy sequences $x = \{x_n\}$, $x_n \in R$, where two sequences $x = \{x_n\}$ and $y = \{y_n\}$ are not considered distinct if $\|x_n - y_n\| \to 0$ and the elements $x \in R$ are identified in $\hat{R}$ with the sequence $\{x,x,x,....\}$. The operations with these sequences are defined by $\alpha x = \{\alpha x_n\}$, $x + y = \{x_n + y_n\}$, $(x,y) = \lim (x_n,y_n)$.

18. That $(N + I)^p$ is e.s.a. can be proven in many ways. It also follows from the fact that $(N + I)^p$ is an elliptic element in the enveloping algebra of a group representation. See Ref. 22.
19. It is well known that P, Q and H and consequently a^+ and a cannot be continuous operators with respecto to τ_H and are, therefore, not defined on the whole Hilbert space.
20. Let $\varphi_n \rightarrow \varphi$, with respect to τ_Φ, i.e., $(\varphi_n - \varphi) \rightarrow 0$, wth respect to τ_Φ. Let us assume that A was not defined on φ. As A is a continuous operator, $A(\varphi_n - \varphi) \rightarrow 0$, with respect to τ_Φ. Therefore, we can define A on φ by
$A\varphi = \tau_\Phi\text{–}\lim A\varphi_n$.
21. The definition of closed and closable operators are given in Appendix III. P,Q and H are closable because they are symmetric and a symmetric operator admits a closure.

22. E. Nelson, W.F. Stinespring, Amer Journ. Math., 81 ,547 (1959).
23. This definition can be seen in I.M. Gel'fand, N.J. Vilenkin, "Generalized Functions", vol IV, Academic Press, New York.
24. J.E. Roberts, Commun. Math. Phys., 3, 98 (1966).
25. A. A. Kirilov, Kokl. Akad, Nauk, SSSR 130, 966 (1976).
26. A. Bohm, Appendix B of J. Math. Phys., 8 ,1557 (1967).
27. B. Nagel, Lecture Notes, College de France (1970).
28. J. Horvath, "Topological Vector Spaces and Distributions", Addison Wesley, Reading, Mass, (1966).
29. M. Reed and B. Simon, "Functional Analysis" ,Academic Press, New York (1970).

30 These mappings are onto, except for that one which maps all the elements in $\Phi^\times$ into zero.
31. See Reference 29.
32. See any text in measure theory, for instance: A. Friedmann, "Foundations of Modern Analysis",
33. J Dixmier, Comp. Math. 13, 263 (1985) and references thereof.
34. If Φ satisfies the first axiom of countability, its topology can be thorougly described by the passage to the limit of countable sequences.
35. This is not the only possible choice of $d(\lambda)$, because $d\mu(\lambda) = \rho(\lambda)d\lambda$, where $\rho(\lambda)$ is an integrable non negative real function on R, will work as well.
36. Usually, a tempered distribution is defined to be a continuous linear functional on S and not antilinear as done here.
37. See for instance G. Bachman and L. Narici, "Functional Analysis", Academic Press, New York (1966).
38. This notation is often used in textbooks on distribution theory. See for instance: Vu Khac Khoan,"Distributions, Analyse de Fourier, Operateurs aux Derivées Partielles". Viubert, Paris (1972), vol 1.
39. G. Lindblad, B. Nagel, Ann. Inst. Poincaré, Sect. A. (N.S.) 13, 27 (1970).

<u>CHAPTER II</u>

I. INTRODUCTION

In a scattering process, one has a particle or beam of particles which move freely toward a region in which forces act on them. This region is called the scattering region. After a while the paticles leave the scattering region. They have been perturbed and move freely again. The scheme depicted on Fig. 1 summarizes this situation.

For the sake of simplicity, we shall assume that only one particle undergoes scattering. What we have in quantum mechanics is an incident state vector and a scattered one. Both are free of interaction and consequently evolve freely. If we represent the incident state vector by f^{in} and the scattered one by g^{out}, we may write

$$g^{out} = Sf^{in} \tag{2-1}$$

S is called the S-operator. It depends on the properties of the scattering region (nature of the forces, their strengths, etc.).

We shall here consider a spinless particle and also assume that the angular momentum is a conserved quantity. Then, we can express the Hilbert space of states H as a direct sum of Hilbert spaces each having a well determined value of the angular momentum

$$H = \bigoplus_{l=0}^{\infty} H_l \tag{2-2}$$

For each of the H_l, we have an expression analogous to (2-1)

$$g_l^{out} = S_l \, f_l^{in} \tag{2-3}$$

f^{in}_l , $g^{out}_l \in H_l$, where l denotes a value of the angular momentum. In the remainder of this presentation, we shall consider the value of the angular momentum to be fixed and omit the label in (2-3) for simplicity.

Let H_o be the hamiltonian for the free particle. In the representation which diagonalizes H_o, we can write (2-3) as

$$g^{out}(E) = S(E)f^{in}(E) \tag{2-4}$$

The variable E is the kinetic energy, i.e., $E = P^2/2m$, where $P = |P|$ is the modulus of the momentum of the free particle. S(E) is here a matrix function of E called the S-matrix (S(E) depends also on the value l of the angular momentum). S(E) in (2-4) is physically defined for positive values of E only. However, it admits an analytic continuation to the two sheeted Riemann surface associated with the transformation $w = z^2$.[1,2] In the sequel, we consider S(E) to be meromorphic (analytic except for isolated singularities) on this Riemann surface and allow it to have one (or more) of the following types of singularities.

1. Simple poles on the negative real axis of the first sheet; these correspond to bound states (which can exist inside the scattering region).

2. Simple poles on the negative real axis of the second sheet; these correspond to the so-called virtual states.

3. Simple poles on the second sheet which are symmetric with respect to the positive real axis (branch cut). The positions of the pairs are denoted by

$$Z_R = E_R \pm i\Gamma/2,\ E_R, \Gamma > 0. \tag{2-5}$$

Each pair of poles corresponds to a resonance with resonant energy E_R and width Γ.

Since S(E) is defined on a Riemann surface which is made up of two planes, we have to distinguish the values on the first sheet from the values on the second, henceforth, they will be denoted as $S_I(\omega)$ and $S_{II}(\omega)$, respectively. From now on, we shall use the greek letter ω to denote the complex variable $\omega = E + ik$, reserving the symbol E for the real part of ω. When ω approaches the cut, the following boundary values hold

$$\lim S_I(E + ik) = \lim S_{II}(E - ik) \quad \text{as } k \to 0 \tag{2-6-A}$$

$$\lim S_I(E - ik) = \lim S_{II}(E + ik) \quad \text{as } k \to 0 \tag{2-6-B}$$

We shall use the following notation

$$S(E \pm io) = \lim S_I(E \pm ik) \quad \text{as } k \to o \tag{2-7}$$

and one can prove that

$$S(E + io) = S^*(E - io) \tag{2-8}$$

where the star stands for complex conjugation.[1,2)]

Those readers who are not familiar with the ideas on scattering theory, can get acquainted with them by reading the later chapters of the book by A. Bohm cited in the first reference of the present chapter. A comprehensible review of the idea and the properties of the S-matrix can be found in Reference 2.

We want to add a few more facts about resonant scattering. Consider a scattering experiment in which a resonance has been produced. This resonance is characterized by the numbers E_R and Γ. After formation, it eventually decays. The whole process can be picturized by means of the diagram on Figure 2.

A free state f^{in} enters the region of interaction where the resonance or semibound state is produced. After a time, which is statistically related to Γ, the resonance decays and is observed far from the region of interaction as a new free state g^{out}.

This process can be split into two parts: The first will account for the formation of the resonance. The second for the decay. Before t = 0 a free state grows into a quasistationary state.[1]

After $t = 0$ the latter decays into a free state. We can draw a diagram to picture each of these parts.

Figure 3 shows a growing process. Here, we ignore the decaying part. On the other hand, figure 4 depicts the decaying process and the growing part is ignored.

II. THE MOLLER WAVE OPERATORS

In this section, we shall assume that the interaction that causes formation or decay of the resonance is given by a potential V. The hamiltonian H is thus a sum of the hamiltonian describing the free particle, H_o plus the perturbation V, $H = H_o + V$. All these operators act on a Hilbert space of states $\mathcal{H}$. H has a continuous spectrum which coincides with R^+, the positive part of the real line, and it can also have a discrete spectrum describing the bound states. We do not admit a singular part in the continuous spectrum. According to the spectral theory,[3-6] we can split $\mathcal{H}$ into two orthogonal subspaces:

$$\mathcal{H} = \mathcal{H}_p \oplus \mathcal{H}_{ac} \qquad (2\text{-}9)$$

where: a) $\mathcal{H}_p$ is the Hilbert space spanned by the eigenvectors belonging to the discrete spectrum of H, and b) $\mathcal{H}_{ac}$ is its orthogonal subspace (see Appendix III). The spaces $\mathcal{H}_p$ and $\mathcal{H}_{ac}$ are invariant under the action of H. This means that if $g_1 \in \mathcal{H}_p$ is a vector such that $Hg_1 \in \mathcal{H}$ then $Hg_1 \in \mathcal{H}_p$ and that $g_2 \in \mathcal{H}_{ac}$ and $Hg_2 \in \mathcal{H}$ implies $Hg_2 \in \mathcal{H}_{ac}$.

We call H_1 the restriction of H to $\mathcal{H}_{ac}$. This means that if $f \in \mathcal{H}_{ac}$ and f is in the domain of H (see Appendix III):

$$Hf = H_1 f \qquad (2\text{-}10)$$

Pick now $f^{in} \in \mathcal{H}$. It behaves like a "free" vector if it evolves under the action of the free dynamics, i.e., $f^{in}(t) = e^{-itH_o} f^{in}$ if $f^{in} = f^{in}(0)$. If we assume that any state in $\mathcal{H}$ can be prepared and evolve freely, f^{in} can be any vector in $\mathcal{H}$. Now, let $f^+ \in \mathcal{H}$ such that it evolves in time according to $e^{-itH} f^+ = f^+(t)$ (again $f^+ = f^+(0)$). If $f^+ \in \mathcal{H}_{ac}$ then $f^+(t) = e^{-itH} f^+ = e^{-itH_1} f^+$.

Assume that, when $t \to -\infty$, before the interaction takes place, f^+ behaves like a free state, say f^{in}. This means that if these two states f^+ and f^{in} evolve toward the past, they must coincide in the limit:

$$\lim \ (f^+(t) - f^{in}(t)) = 0 \ ; \ t \to -\infty \qquad (2\text{-}11)$$

or, more precisely

$$\lim \ \|f^+(t) - f^{in}(t)\| = 0 \ ; \ t \to -\infty \qquad (2\text{-}12)$$

This limit can be written as a function of the values of the state vectors at zero.

$$\lim \ \|e^{-itH} f^+ - e^{-it H_o} f^{in}\| = 0 \ ; \ t \to -\infty \qquad (2\text{-}13)$$

According to the law of quantum mechanical time development, e^{-itH} and its inverse e^{itH} are unitary operators. Unitary operators preserve norms (Appendix III). Thus

$$\|e^{-itH} f^+ - e^{-itH_o} f^{in}\| = \|e^{itH} e^{-itH} f^+ - e^{itH} e^{itH_o} f^{in}\|$$

$$= \|f^+ - e^{itH} e^{-itH_o} f^{in}\| \qquad (2\text{-}14)$$

Hence

$$\lim \ \|f^+ - e^{itH} e^{-itH_o} f^{in}\| = 0 \ ; \ t \to -\infty \qquad (2\text{-}15)$$

This limit may or may not exist and its existence depends on the hamiltonian H. If we assume that (2-15) is valid for any $f^{in} \in \mathcal{H}$, we can write:

$$\lim \quad e^{itH}e^{-itH_o}f^{in} = f^+ ,\ \forall f^{in} \in H . \tag{2-16}$$

and an operator Ω^+ by:

$$\Omega^+ f^{in} = \lim \quad e^{itH} e^{-itH_o}f^{in} = f^+ \ ; t \to +\infty \tag{2-17}$$

Ω^+ is an isometric mapping on H . If we write $\boldsymbol{H_+} = \Omega^+ H$, one can prove that Ω^+ is a unitary mapping from H onto $\boldsymbol{H_+}$ (see Appendix III for the definition of isometric and unitary mappings).

Now, let g^{out} and g^- be vectors in H . g^{out} evolves freely (its time development is given by H_o) and g^- evolves according to the time development given by H. We assume that in the limit when $t \to +\infty$, when the interaction is not effective any longer, we shall not be able to distinguish between g^{out} and g^-. In that case we say g^- is asymptotically free when $t \to +\infty$. We write

$$\lim \quad \|e^{-itH} g^- - e^{-itH_o}g^{out}\| = 0 \ ; \ t \to \infty \tag{2-18}$$

which is equivalent to

$$\lim \quad \|g^- - e^{itH} e^{-itH_o}g^{out}\| = 0 \ ; t \to \infty \tag{2-19}$$

Again, this limit may or may not exist. If we can guarantee its existence for any $g^{out} \in H$, we can define an operator Ω^- by

$$\Omega^- g^{out} = \lim \quad e^{itH} e^{-itH_o}g^{out} = g^- \ ; \ t \to \infty \tag{2-20}$$

Ω^- is an isometry on H and it is a unitary mapping from H onto $\boldsymbol{H_-} = \Omega^- H$.

We call Ω^+ and Ω^- the Moller wave operators.

In physical terms, f^+ is an interacting state prepared as f^{out} in the remote past before the interaction was effective. g^- is an interacting state vector which can be observed as g^{out} in the distant future when the interaction V ceases to be effective.

A scattering process given by the interaction $V = H - H_o$ is called asymptotically complete if

$$\boldsymbol{H_+} = \boldsymbol{H_-} = H_{ac} \tag{2-21}$$

In terms of physics, this means that for any free {incoming/outgoing} state we shall have a free {outgoing/incoming} state so that no free state becomes bound after being captured by the scattering forces and no bound state can become free.

As a consequence of asymptotic completeness, $\Omega^\pm$ are unitary mappings from H onto H_{ac}. The following results can also be proven

$$H_o = (\Omega^\pm)^{-1} H_1 \Omega^\pm ;\ H_1 = \Omega^\pm H_o (\Omega^\pm)^{-1} \tag{2-22}$$

$$\Omega^\pm e^{-itH_o} \ = e^{-itH_1}\Omega^\pm \tag{2-23-A}$$

$$e^{-itH_1} \ = \Omega^\pm e^{-itH_o}(\Omega^\pm)^{-1};\ e^{-itH_o} = (\Omega^\pm)^{-1} e^{-itH_1} \ \Omega^\pm \tag{2-23-B}$$

We now assume that a free state is prepared in the remote past as f^{in}, undergoes the action of the scattering forces and evolves as $f^+(t)$ if $t < 0$ and as $g^-(t)$ if $t > 0$ ($f^+(o) = g^-(o)$) and is finally observed in the distant future as the free state g^{out}. Then relation (2-1) remains valid here: $g^{out} = Sf^{in}$. S relates the prepared and the observed free states; therefore, S can exist if no Moller operators $\Omega^\pm$ existed, then one could show that[3,5)]

$$S = (\Omega^-)^{-1} \Omega^+ \tag{2-25}$$

Furthermore, if we have asymptotic completeness, S is unitary.

III. THE HARDY CLASS FUNCTIONS ON A HALF PLANE

In this section, we shall briefly review the properties of the Hardy class functions on a half plane. We shall extensively use these functions in our description of resonances. Later, we shall construct rigged Hilbert spaces $\Phi \subset H \subset \Phi^{x}$ such that the vectors on Φ can be realized as Hardy class functions on a half plane. From now on, we shall call C^{+} the set of complex numbers such that its imaginary part is equal or bigger than zero, i.e., the upper half plane plus the real line. C^{-} will be the lower half plane plus the real line, i.e., the set of complex numbers with negative or zero imaginary part.

Definition: Let f(x) be a complex function on R, i.e., a function from R into C. f(x) is a Hardy class function on the upper half plane C^{+} if:

1. There exists a complex value function f(z) analytic for Imz > 0, i.e., on C^{+} minus the real line.

2. For any fixed $y_o > 0$, consider the straight line $z = x + iy_o$. Then $f(x + iy_o)$ is a complex function of the real variable x. We want this function to be square integrable, i.e.,

$$\int_{-\infty}^{\infty} |f(x + iy_o)|^2 \, dx < \infty \tag{2-25-A}$$

Furthermore, all the integrals in (2-25-A) have to be bounded by the same positive constant K. Thus,

$$\sup_{\operatorname{Im} z = y > 0} \int_{-\infty}^{\infty} |f(x + iy)|^2 \, dx < K < \infty \tag{2-26-A}$$

independently of the fixed value of y > 0 we have chosen for each integral.

3. The complex function of real variable f(x) gives the boundary values of f(z) on R. In other words

$$\lim f(x + iy) = f(x) \; ; \; y \to 0 \tag{2-27-A}$$

(2-27-A) is valid for almost every $x \in R$. This means that (2-27-A) is true for every $x \in R$ except possibly for a set of real numbers with zero Lebesgue measure. Equivalently, we say that (2-27-A) is valid almost everywhere on R.

As a consequence of the definition we have

$$\int_{-\infty}^{\infty} |f(x)|^2 dx \quad \leq \; K$$

(2-28-A)

which means that the boundary function f(x) is also square integrable. Therefore, to any Hardy class function we associate a complex function analytic on the open set {Im z > 0} which fixes f(x) almost everywhere on R. Later, we shall see that the reciprocal is also true and that we can recover the values of f(z) from those of f(x). For this reason, we shall use the term "Hardy class function" for f(x) as well as for f(z), although we mostly apply it for f(x).

After (2-28-A), we see that a Hardy class function on the upper half plane is square integrable, i.e., it belongs to $L^2(R)$. Hereafter, we shall call H_+ the space of the functions fulfilling 1-3. H_+ is a linear subspace of $L^2(R)$. If we endow H_+ with the topology inherited from $L^2(R)$, H_+ is a closed subspace of $L^2(R)$ and hence a Hilbert space.

In analogy to the definition of H_+, we can also define a Hardy class function g(x) on the lower half plane. We call the set of these functions H_-. A complex function of R, g(x) belongs to H_- if:

1. There exists a complex valued function g(z) analytic on Im z < 0.

2. For every $y_o < 0$, the complex function of the real variable $g(x + iy_o)$ is square integrable, i.e.,

$$\int_{-\infty}^{\infty} |g(x + iy_o|^2 dx \; < \infty$$

(2-25-B)

and we can find a positive constant K_l larger than the integrals (2-25-B) for any value of y < 0:

$$\sup_{y<0} \int_{-\infty}^{\infty} |g(x + iy)|^2 dx \; < K_1 < \infty$$

(2-26-B)

3. The complex function of real variable g(x) gives the boundary values of g(z) on R. This implies that

$$\lim_{y \to 0} g(x + iy) = g(x) \tag{2-27-B}$$

for almost every $x \in R$.

As a consequence of the definition, one obtains:

$$\int_{-\infty}^{\infty} | g(x) |^2 \, dx \leq K_1 \tag{2-28-B}$$

which implies that $H_- \subset L^2(R)$. H_- is also a closed linear subspace of $L^2(R)$ and hence a Hilbert space (we define on H_- as well as on H_+ the scalar product these spaces have as subspaces of $L^2(R)$).

Properties of the Hardy class functions

Let f(x) be a complex function of the real variable. Its Hilbert transform $\mathbb{H}$ is defined by

$$\mathbb{H}\, f(t) = \qquad (2\text{-}29)$$

$$= \pi^{-1}\, P \int_{-\infty}^{+\infty} \frac{f(x)}{x - t}\, dx$$

where P stands for the Cauchy principal value.[7]

$\mathbb{H}$ is obviously linear. Furthermore, if f(x) is in $L^2(R)$, $\mathbb{H}$ f(x) is also in $L^2(R)$.[8]

Any $f(x) \in H_{\pm}$[9] can be decomposed in its real and imaginary parts. f(x) = h(x) + ig(x) where h(x) and g(x) are real functions on R. If $f(x) \in H_+$, a result by Titchmarsh,[8] states that

$$\mathbb{H} h(x) = g(x) \quad \text{and} \quad \mathbb{H}\, g(x) = -h(x) \qquad (2\text{-}30\text{-}A)$$

The reciprocal is also true. If h(x) and g(x) are real functions and fulfill (2-30-A), then $f(x) = h(x) + ig(x) \in H_+$.[10]

If now, $f(x) \in H_-$ a similar result says that

$$\mathbb{H} h(x) = -g(x) \quad \text{and} \quad \mathbb{H} g(x) = h(x) \qquad (2\text{-}30\text{-}B)$$

The converse is true as well: if g(x) and h(x) are real and fulfill (2-30-B), f(x) belongs to H_+

> The proof is very easy again: f(x) = h(x) + ig(x) and $f^*(x) = h(x) - ig(x)$. If $f(x) \in H_+$, $\mathbb{H}$ h(x) = g(x) and $\mathbb{H}$ g(x) = -h(x). Call $\nu(x) = -g(x)$. Then $f^*(x) = h(x) + i\nu(x)$ and $\mathbb{H}\, \nu(x) = h(x)$ by the linearity of $\mathbb{H}$. These two later equations assert that $f^* \in H_-$ because of the reciprocal of (2-30-B). The converse is proven analogously.

The second result establishes that if f(x) is a real function, it cannot belong to either H_+ or H_-, unless it is the zero element. For if $f(x) = h(x) \in H_+$, then $f^*(x) = h(x) \in H_-$ and consequently it must be the zero function.

Titchmarsh has also proven a result which allows us to recover the values on the half plane for any function in H_+ or H_-, provided we know its values on the real line. Since we shall need it in other sections, we label it as Theorem 1 or the Titchmarsh theorem.

Theorem 1 (Titchmarsh).[10,12,13] Let f(x) be a function in H_+. Then for all z = x + iy with y > 0, we can write

$$f(z) = (2\pi i)^{-1} \int_{-\infty}^{\infty} \frac{f(x)\,dx}{x - z}$$

(2-31-A)

and the integral vanishes if $y < 0$. Conversely, if f(x) is a square integrable function and the integral in the right hand side of (2-31-A) vanishes for $y < 0$, then $f(x) \in H_+$ and (2-31-A) holds.

An analogous result can be obtained for $f(x) \in H_-$. If $y < 0$, we can write

$$f(z) = (-2\pi i)^{-1} \int_{-\infty}^{\infty} \frac{f(x)\,dx}{x - z}$$

(2-31-B)

This equation is valid for any z with negative non zero imaginary part. The integral is zero for the values of z with $y > 0$. Conversely, if $f(x) \in L^2(R)$ and the integral in the right hand side of (2-31-B) vanishes for $y > 0$, then $f(x) \in H_-$ and (2-31-B) holds.

It is important to stress that {(2-31-A) / (2-31-B)} gives the values of a function in $\{H_+ / H_-\}$ on $\{C^+ - R / C^- - R\}$. This formula is valid for $\{y > 0 / y < 0\}$ only. It does not prohibit a possible analytic continuation of f(z) into {the lower half plane / the upper half plane}.

The Fourier Transform

In this subsection we wish to recall some important properties of the Fourier transform. The Fourier transform is a powerful tool for the theory of the Hardy class functions, as well as an important technique in theoretical and mathematical physics.

Let $f(x) \in L^2(R)$, its Fourier transform is defined by

$$\hat{f}(t) = (\mathcal{F} f)(t) = \frac{1}{(2\pi)^{1/2}} \int_{-\infty}^{+\infty} e^{-itx} f(x)\,dx$$

(2-32-A)

Therefore, the Fourier transform is a unitary mapping from $L^2(R)$ onto itself. This statement is known in the literature as the Plancherel theorem. Therefore, $\|\mathcal{F} f\| = \|f\|$ and $\mathcal{F}$ has an inverse on $L^2(R)$, $\mathcal{F}^{-1}$, which is given by

$$\check{f}(x) = (\mathcal{F}^{-1} f)(x) = \frac{1}{(2\pi)^{1/2}} \int_{-\infty}^{+\infty} e^{itx} f(t)\, dt$$

(2-32-B)

Therefore, the Fourier transform $\hat{f}(t)$ of f(x) as well as its inverse Fourier transform, $\check{f}(x)$, are well defined square integrable functions. On the other hand, the Schwartz space S is a linear subspace of $L^2(R)$. Consequently, the Fourier transform and the inverse Fourier transform of f(x) $\in$ S are well defined and square integrable. As a matter of fact, one can prove much more:

1. If f(x) $\in$ S, $\mathcal{F}$ f and $\mathcal{F}^{-1}$ f belong to S.
2. $\mathcal{F}$ and $\mathcal{F}^{-1}$ are bijections from S onto itself. A bijection is a one to one onto mapping.
3. $\mathcal{F}$ and $\mathcal{F}^{-1}$ are continuous mappings (and hence bicontinuous) on S provided we endow S with its topology as defined in Section 4 of Chapter I.

$\mathcal{F}$ can be extended to S^x and this extension can be made as follows: First, note that since $\mathcal{F}^* = \mathcal{F}^{-1}$, $\mathcal{F}^*$ leaves S invariant. Then, if F $\in$ S^x

$$\langle \mathcal{F}^* f | F\rangle = \langle f | \mathcal{F}^x F\rangle \qquad \forall f \in S \qquad (2\text{-}33\text{-}A)$$

An equivalent form is

$$\langle \mathcal{F} f | \mathcal{F}^x F\rangle = \langle f | F\rangle \qquad \forall f \in S \qquad (2\text{-}33\text{-}B)$$

$\mathcal{F}^x$ is really an extension of $\mathcal{F}$, for if F $\in$ $L^2(R)$

$$\langle \mathcal{F} f | \mathcal{F}^x F\rangle = \langle f | F\rangle = (f, F) = (\mathcal{F} f, \mathcal{F} F) \qquad (2\text{-}34)$$

The latter identity follows from the unitarity of $\mathcal{F}$. The right hand side of (2-34) is equal to $\langle \mathcal{F} f | \mathcal{F} F\rangle$. Thus,

$$\langle \mathcal{F} f | \mathcal{F}^x F\rangle = \langle \mathcal{F} f | \mathcal{F} F\rangle \qquad (2\text{-}35)$$

for all f $\in$ S and F $\in$ $L^2(R)$. Since $\mathcal{F}$ is onto, $\mathcal{F}$ f represents an arbitrary element of S. Therefore, $\mathcal{F}^x F = \mathcal{F} F$, for all F $\in$ $L^2(R)$.

In the same manner, we can extend $\mathcal{F}^{-1}$ to S^x. Its extension, $(\mathcal{F}^{-1})^x$, as well as $\mathcal{F}^x$ are obviously linear on S. Moreover, one can also show:

1. $\mathcal{F}^x$ F and $(\mathcal{F}^{-1})^x$ F for F $\in$ S^x are continuous antilinear functionals on S and, therefore, belong to S^x (continuity of functionals has already been defined in Chapter II, Section 3).
2. $\mathcal{F}^x$ and $(\mathcal{F}^{-1})^x$ are bijections from S^x onto itself.
3. We can endow S^x with the weak topology (see Chapter I). With respect to this topology, $\mathcal{F}^x$ and $(\mathcal{F}^{-1})^x$ are bicontinuous. (They and their inverses are continuous). This means, in particular, that if a sequence $(F_n) \in S^x$ converges to F $\in$ S^x, then $\mathcal{F}^x F_n \to \mathcal{F}^x F$ and $(\mathcal{F}^{-1})^x F_n \to (\mathcal{F}^{-1})^x F$.
4. $(\mathcal{F}^{-1})^x$ is the inverse of $\mathcal{F}^x$.

The proof of this statement is easy:

$$\langle \mathcal{F} f | F\rangle = \langle f | (\mathcal{F}^*)^x F\rangle = \langle f | (\mathcal{F}^{-1})^x F\rangle \qquad f \in S,\ F \in S^x \qquad (2\text{-}36)$$

On the other hand

$$\langle f \mid (\mathcal{F}^{-1})^{x} \mathcal{F}^{x} F\rangle = \langle \mathcal{F}\; f \mid \mathcal{F}^{x} F\rangle = \langle \mathcal{F}^{*} \mathcal{F}\; f \mid F\rangle$$
$$= \langle \mathcal{F}^{-1} \mathcal{F}\; f \mid F\rangle = \langle f \mid F\rangle \qquad f \in S,\; F \in S^{x} \qquad (2\text{-}37)$$

Therefore, $(\mathcal{F}^{-1})^{x}\mathcal{F}^{x}F = F$. $F \in S^{x}$ which implies

$$(\mathcal{F}^{-1})^{x} \mathcal{F}^{x} = I \qquad (2\text{-}38\text{-}A)$$

where I is the identity of S^{x}. Since $\mathcal{F}^{x}$ is bijective, it has an inverse and (2-38-A) yields

$$(\mathcal{F}^{-1})^{x} = (\mathcal{F}^{x})^{-1} \qquad (2\text{-}38\text{-}B)$$

The Fundamental Theorems

In this subsections, we need to present the most important results we need for the construction of certain spaces of Hardy class functions. We shall make later use of these results in order to describe resonances and virtual states (see Chapters III and IV). We label these results here as Theorem II, III and IV. We do not give proofs, since they involve techniques which are outside of the scope of these lectures.

Definition: We identify two functions which are different on a set of zero Lebesgue measure only. Therefore, $L^2(R^+)$ and $L^2(R^-)$ are, properly speaking, spaces of classes of functions as well as $L^2(R)$. The space $L^2(R^+)$ is the set of all the complex functions on the positive real line which are also square integrable, i.e. $f(x) \in L^2(R^+)$ iff [14)]

a) $f(x) : R^+ \to C$, i.e. f(x) is a complex function on the positive part of the real line.

b) $\int_0^{\infty} |f(x)|^2\, dx < \infty$

Similarly, g(x) belongs to $L^2(R^-)$ iff

a) $g(x) : R^- \to C$, where R^- is the negative part of the real line.

b) $\int_{-\infty}^{0} |g(x)|\, dx < \infty$

Any function $f(x) \in L^2(R^+)$ can be extended into R^- to be in $L^2(R)$ if we define $f(x) = 0$ for the negative values of x. Also $g(x) \in L^2(R^-)$ can be extended into R^+ to be in $L^2(R)$ if we state that $g(x) = 0$ if $x \in R^+$. In order to simplify the notation, we shall always call $L^2(R^+)$ and $L^2(R^-)$ the spaces of these functions respectively, whether extended or not. $L^2(R^+)$ and $L^2(R^-)$ are Hilbert spaces. Their scalar products are defined as

$$(f,g) = \int_0^\infty f^*(x)\, g(x)\, dx \quad \text{on} \quad L^2(R^+)$$

$$(h,s) = \int_{-\infty}^0 h^*(x)\, s(x)\, dx \quad \text{on} \quad L^2(R^-)$$

(2-39)

After the above considerations and the definition of direct sum given in Appendix III, we have

$$L^2(R) = L^2(R^+) \oplus L^2(R^-) \qquad (2\text{-}40)$$

Now, we can present the first of the above mentioned results:

<u>Theorem II (Paley-Wiener I)</u>[12,13,15,16])

a) The Fourier transform F is a bijection from $L^2(R^-)$ onto H_+ and also from $L^2(R^+)$ onto H_-.

This theorem gives us a method to construct H_+ and H_- by takfing the Fourier transform of all functions in L^2 (R^-) and in $L^2(R^+)$ respectively (or the inverse Fourier transform of the functions in $L^2(R^+)$ and in $L^2(R^-)$ respectively). It also answers the question: when is a function $f(x) \in L^2(R)$ an element of the spaces $H_\pm$? The answer is: if $F\,f \in L^2(R^+)$ or $F^{-1} f \in L^2(R^-)$, then f(x) belongs to H_+; if $F\,f \in L^2(R^-)$ or $F^{-1} f \in L^2(R^+)$, then f(x) belongs to H_-. Otherwise, f(x) is neither in H_+ nor in H_-. Note that (2-30) also solves this problem.

On the other hand, $H_\pm$ are closed subspaces of $L^2(R)$ and, therefore, they are Hilbert spaces when the scalar product on $H_\pm$ is the one inherited from $L^2(R)$.

> The proof is straightforward: Let (f_n) be a Cauchy sequence of vectors in H_+. It converges to a function $f \in L^2(R)$, because $L^2(R)$ is a Hilbert space. Since the Fourier transform is a continuous mapping from $L^2(R)$ onto itself, $F\,f_n \to F\,f$ in $L^2(R)$. Theorem II in its part b) says that $F\,f_n \in L^2(R^+)$ which is a Hilbert space. Since $(F\ f_n)$ converges in $L^2(R)$ it also converges in $L^2(R^+)$ and its limit $F\ f$ must belong to $L^2(R^+)$. Let us apply now part b) of Theorem II again. $F^{-1}(F\,f) = f \in H_+$. We have proven that any Cauchy sequence of vectors in H_+ has a limit in H_+,

which proves that H_+ is closed. Similar arguments are valid for H_- as well.

A second remark has its origin in the following idea: F and F^{-1} are unitary mappings on $L^2(R)$, i.e., they are bijective and norm preserving. Thus, Theorem II can be restated as follows:

b) The inverse Fourier transform F^{-1} is a unitary mapping from $L^2(R^+)$ onto H_+ and also from $L^2(R^-)$ onto H_-.

An important consequence of Theorem II is

$$L^2(R) = H_+ \oplus H_- \qquad (2\text{-}41)$$

The proof is quite simple. According to (2-40) any $f \in L^2(R)$ can be written as $f = f_1 + f_2$ where $f_1 \in L^2(R^-)$ and $f_2 \in L^2(R^+)$. For any $g \in L^2(R)$ there exists $f \in L^2(R)$ such that $g = F\ f = F\ f_1 + F\ f_2$. $F\ f_1 \in H_+$ and $F\ f_2 \in H_-$. Thus, any function in $L^2(R)$ can be written as a sum of one in H_+ and another one in H_-. Since $H_+ \cap H_- = \{0\}$, the zero element, this sum is direct and ((2-41) follows (see Appendix 2).

Take now $f(x) \in H_+$. Then $f^*(x) \in H_-$ and vice versa. Therefore, $f(x)$ and $f^*(x)$ are orthogonal (see (2-41)). Furthermore,

$$0 = (f^*,f) = \int_{-\infty}^{\infty} f(x)\, f(x)\, dx = \int_{-\infty}^{\infty} f^2(x)\, dx \qquad (2\text{-}42)$$

This property is also valid for any $g(x) \in H_-$.

Finally, if we call F^2 the action of two successive Fourier transforms and F^{-2} the action of two successive inverse Fourier transforms, a combined use of a) and b) yields:

$$F^2 L^2(R^-) = L^2(R^+) \qquad (2\text{-}43\text{-}A)$$

$$F^{-2} L^2(R^+) = L^2(R^-) \qquad (2\text{-}43\text{-}B)$$

The next result will answer the question: Is a function in $H_\pm$ determined by its values on R^+? An affirmative answer should provide a method to recover the values $f(z)$ on $C^\pm$ of the function from its values on R^+. In particular w should be able to obtain the values of $f(x)$ for $x \in R^-$ from those for $x \in R^+$. This is possible and can be done with the help of the Mellin transform.

<u>Definition</u>: Let $h(x)$ be a function on R^+. Its Mellin transform is a function on R (the whole real line!) defined as:

$$E\,(s) = \frac{1}{(2\pi)^{1/2}} \int_0^{\infty} h(x)\, x^{is-1/2}\, dx \qquad (2\text{-}44)$$

provided that this integral exists for almost all $s \in R$.

Theorem III (van Winter)[17]

A function $h(x) \in L^2(R^+)$ can be extended to R^- to become a function in H_+ if and only if its Mellin transform (2-44) satisfies

$$\int_{-\infty}^{\infty} (1 + e^{2\pi s}) |E(s)|^2 ds < \infty \tag{2-45}$$

This extension is unique. The values of h(z) for $z = re^{i\Theta} : 0 \leq \Theta \leq \pi, r > 0$ are given by

$$h(re^{i\Theta}) = \frac{1}{(2\pi)^{1/2}} \int_{-\infty}^{\infty} E(s) (re^{i\Theta})^{-is-1/2} ds \tag{2-46}$$

In particular, for negative values of x, h(x) is given by

$$h(-x) = \frac{1}{(2\pi)^{1/2}} \int_{-\infty}^{\infty} E(s) (xe^{i\pi})^{-is-1/2} \tag{2-47}$$

(2-46) is the inverse Mellin transform of $E(s)$. For positive values of x, h(x) is our original function. We do not want to discuss here these properties of the Mellin transform and we refer the reader to the classical literature on the subject.[18]

A similar result can be obtained for H_-. In fact, if the complex conjugate of $g(x) \in L^2(R^+)$, $g^*(x)$, satisfies the premises of the van Winter theorem, then it can be uniquely extended to R^- in such a way that the extension belongs to H_+. In this case, the complex numbers g(x) represent the values on R^+ of a function in H_-. Reciprocally, if $g^*(x)$ does not satisfy the requirement of Theorem II, g(x) cannot have an extension in H_-.

For proof, details and a deeper insight on the van Winter theorem see References 17 and 19 at the end of this chapter.

The next theorem[20] is useful for the description of virtual states as we shall see in Chapter IV, Section 3.

Theorem IV. Let f(t) be in $L^2(R)$. Then $e^{b|t|} f(t) \in L^2(R)$ for $b < a$ if and only if $(F f)(x)$ has an analytic continuation into the strip $\{|\text{Im } z| < a\}$ such that

1.

$$\int_{-\infty}^{\infty} |(\mathcal{F}\ f)(x+iy)|^2\, dx < \infty$$

(2-48)

provided $|y| = |\text{Im } z| < a$

2. For any $b < a$

$$\sup_{|y|=b} \int_{-\infty}^{\infty} |(\mathcal{F}\ f)(x+iy)|^2\, dx < \infty$$

(2-49)

1. gives us the following information: Let f(t) be a square integrable function such that $e^{b|t|}f(t)$ is also square integrable for any positive b smaller than a fixed number a. Then, the Fourier transform of f(t), $(\mathcal{F}\ f)(x)$, is a function of complex variable analytic on the strip $\{z \in C$ such that $|y| < a$, $z = x + iy\}$. The converse is also true. For any straight line $z = x + iy_0$ with $-a < y_0 < a$, the complex function of the real variable x, $(\mathcal{F}\ f)(x + iy_0)$, is square integrable.

2. asserts that all the integrals in (2-48) have the same bound if $-b \leq y \leq b$. The positive number b is smaller than a ($b < a$) but otherwise arbitrarily chosen.

For the sake of completeness, we give another result here. It was used to describe virtual states in a previous paper.[21)] This result is very nice as it provides us with an example of a rigged Hilbert space in which the space Φ consists of entire analytic functions.[20)]

Consider the functions in S such that they are different from zero on an arbitrary bounded set. The class of these functions is a linear space called K. Obviously $K \subset S$ and consequently $f \in S$ for $f \in K$. Theorem V below gives further information on the space of the Fourier transforms of the functions in K.

Let f(z) be an entire analytic function (i.e. analytic for all $z \in C$). f(z) is called a function of exponential type if there exist positive constants C_N (depending on N) and ρ such that

$$|f(z)| \leq C_N (1 + |z|)^{-N} e^{\rho |\text{Im } z|} \tag{2-50}$$

<u>Theorem V (Paley-Wiener II)</u>[22,23)] - Let $f \in K$ be zero outside the interval $[-a, a]$. Its Fourier transform $\mathcal{F}\ f$ is the restriction to R of an entire analytic function of exponential type such that $\rho = a$. Reciprocally, the inverse Fourier transform of any entire function of exponential type is a function in K.

The space of entire analytic functions of exponential type is called Z. Thus, the Fourier transform is a bijection from K onto $Z \subset S \subset L^2(R)$. On the other hand K and Z have their own topology[22,23] which do not fulfill the first axiom of countability. The Fourier transform is a bicontinuous bijection from K onto Z, when these topologies are considered. K and Z are also complete nuclear spaces (in some general sense[24]) and their respective topologies are finer than the one they have as subspaces of $L^2(R)$. Therefore, the triplets

$$K \subset L^2(R) \subset K^x \tag{2-51}$$

and

$$Z \subset L^2(R) \subset Z^x \tag{2-52}$$

are rigged Hilbert spaces. The formula

$$\langle \mathcal{F} f \,|\, \mathcal{F}^x F\rangle = \langle f \,|\, F\rangle \qquad \forall f \in K, \quad F \in K^x \tag{2-53}$$

defines the extension of the Fourier transform to the duals: $\mathcal{F}^x : K^x \to Z^x$. $\mathcal{F}^x$ is a bicontinuous bijection from K^x onto Z^x.[22,23]

In the standard literature one calls any continuous linear functional on K a distribution. The linear space of distributions is homeomorphic to K^x. Also the standard literature calls any continuous linear functional on Z an ultradistribution.

References for Chapter II

1. See for details A. Bohm, Quantum Mechanics, Foundations and Applications, Springer, New York (1986), Chapter XVIII.
2. H. M. Nussenzveig, Causality and Dispersion Relations, Academic Press, New York and London (1972).
3. W. Amrein, T. Jauch and B. Sinha, Scattering Theory in Quantum Mechanics, Benjamin, Reading, Mass. (1977).
4. B. Simon, Quantum Mechanics for Hamiltonians Defined as Quadratic Forms, Princeton University Press (1971).
5. M. Reed, B. Simon, Scattering Theory, Academic Press (1977).

6. T. R. Taylor, Scattering Theory, Wiley, New York (1971).

7. For a definition of the Cauchy principal value, see for instance R. V. Churchill, J. W. Brown, R. F. Verhey, Complex Variables and Applications, McGraw Hill, New York (1976). J. R. Marsden, Basic Complex Analysis, Freeman (1973).

8. E. C. Titchmarsh, Introduction to the Theory of Fourier Integrals, Clarendon Press, Oxford (1937).

9. $f(x) \in H_{\pm}$ means that f(x) belongs to either H_+ or H_-. Also here, we shall use the notation $f_{\pm}(x) \in H_{\pm}$. This means that $f_+ \in H_+$ and $f_- \in H_-$.

10. (2-30) are true almost everywhere, i.e., on any point of R, except on a subset of R with Lebesgue measure zero.

11. For the complex conjugate of f(x), we use $f^*(x)$ or $\bar{f}(x)$.

12. P. L. Duren, Theory of H^p Spaces, Academic Press, New York (1970).

13. P. Koosis, Introduction to H^p Spaces, Cambridge University Press (1980).

14. Note that the scalar product on $L^2(R^{\pm})$ is the one inherited from $L^2(R)$.

15. R. Paley and N. Wiener, Fourier Transforms in the Complex Domain, A.M.S., New York (1934).

16. K. Hoffman, Banach Spaces of Analytic Functions, Prentice Hall, New Jersey (1962).

17. C. van Winter, Trans. Am. Math. Soc., 162, 103 (1971).

18. S. Colombo, T. Lavoine, Transformations de Laplace et de Mellin, Gauthier Villars, Paris (1972).

19. C. van Winter, J. Math. An. 47, 633 (1974).

20. M. Reed, B. Simon, Fourier Analysis Self Adjointness, Academic Press, New York (1975).

21. M. Gadella, J. Math. Phys., 24, 2142 (1983).

22. J. M. Gelfand, N. Y. Vilenkin, Generalized Functions, vol. IV, Academic Press, New York (1964).

23. W. Rudin, Functional Analysis, McGraw Hill, New York (1970).

24. K. Maurin, General Eigenfunction Expansion and Unitary Representations of Topological Groups, Polish Scientific Publishers, Warzsawa (1968).

CHAPTER III

I. RIGGED HILBERT SPACES OF HARDY CLASS FUNCTIONS

In this chapter, we shall construct the spaces we need for the description of resonances and virtual states. As mentioned in Chapter II, we assume that these objects are produced in a resonant scattering process, which is described by a hamiltonian $H = H_o + V$, where V is a spherically symmetric potential. H_o is the hamiltonian describing the behavior of the interaction free particle. We call H_o the free hamiltonian. Since H_o is spherically symmetric, it leaves all the proper subspaces of L^2 invariant, where L^2 is the square of the orbital angular momentum. (Here we assume the effects of spin negligible.) If we call $\mathcal{H}_l$ the closure of the subspaces of the eigenvectors of L^2 with eigenvalue $(2l+1)\, h^2$, $\mathcal{H}_o$ is the Hilbert space of the s-states or states with zero angular momentum. From now on, we shall restrict ourselves to these states and construct a theory for resonances and virtual states with zero angular momentum. In order to acomplish that, we need a rigged Hilbert space (R.H.S.) $\Phi \subset \mathcal{H}_o \subset \Phi^x$ with the following properties

1. $H_o\Phi \subset \Phi$
2. H_o is e.s.a. on Φ.[1)]
3. H_o is a continuous operator on Φ.

There are many triplets $\Phi \subset \mathcal{H}_o \subset \Phi^x$ which being given H_o fulfill these conditions. We shall construct two of them which are particularly suited for the description of resonances that we want to discuss here.

For any of these triplets, the Gelfand-Maurin theorem (see Chapter I) will be satisfied. Therefore, there exists a realization $L^2(\Lambda, d\mu(\lambda))$ of the Hilbert space $\mathcal{H}_o$ and a unitary operator.

$U : \mathcal{H}_o \to L^2(\Lambda, d\mu(\lambda))$

with the properties discussed in Chapter I. Λ is here the spectrum of H_o, $\sigma(H_o)$, of which we know[2)] is $\sigma(H_o) = R^+$ and $d\mu(\lambda) = d\mu(E) = dE$. (The points in the spectrum of a Hamiltonian are values of the energy. Thus, we call E to λ. On the other hand, H_o has on $\mathcal{H}_o$ an absolutely continuous spectrum from 0 to ∞.) Thus

$$L^2[\sigma(H_o), d\mu(E)] = L^2(R^+, dE) = L^2(R^+) \tag{3-1}$$

$U\Phi$ is then the realization of the space Φ by a space of functions we shall call D. In order to construct Φ, we have to construct D first. Since H_o is non-degenerate on $\mathcal{H}_o$, it is the multiplication operator on D[3)]:

$$\hat{E} = U\, H_o\, U^{-1} \tag{3-2}$$

Here, we call H_o to the restriction of the interaction-free operator to $\mathcal{H}_o$ as well, and shall do it in the sequel unless otherwise stated.

Since U is unitary, $\hat{E}$ has on D the same properties as H_o on Φ. In particular

1. $\hat{E}D \subset D$
2. $\hat{E}$ is e.s.a. on D
3. $\hat{E}$ is continuous on D.

We have now to find a RHS $D \subset L^2(R^+) \subset D^x$ such that $\hat{E}$ has the above properties. Any function $\varphi(E) \in D$ has the form:

$$U\varphi = \varphi(E) = \langle E|\varphi\rangle = \langle\varphi|E\rangle^* \qquad (= \overline{\langle\varphi|E\rangle}) \tag{3-3}$$

where $\varphi \in \Phi$ and $|E\rangle$ is the generalized eigenvector of H_o with eigenvalue E:

$$(H_o)^x\,|E\rangle = E|E\rangle \tag{3-4}$$

$\hat{E}$ acts on $\varphi(E) \in D$ as a multiplication operator, i.e., if $\varphi(E) \in D$

$$\hat{E}\varphi(E) = E\varphi(E) \qquad E \in R^+ \tag{3-5}$$

To prove this assertion, write $|E\rangle = F_E$ and

$$\langle H_o\varphi|F_E\rangle = \langle\varphi|(H_o)^x F_E\rangle = E\langle\varphi|F_E\rangle = E\langle U\varphi|UF_E\rangle \tag{3-6}$$

$$\langle H_o\varphi|F_E\rangle = \langle UH_o\varphi|UF_E\rangle = \langle UH_oU^{-1}U\varphi|UF_E\rangle \tag{3-7}$$

Therefore,

$$\langle UH_oU^{-1}U\varphi|UF_E\rangle = E\langle U\varphi|UF_E\rangle \tag{3-8}$$

Since UF_E is an arbitrary element of D^x

$$UH_oU^{-1}U\varphi = EU\varphi \tag{3-9}$$

or, since $U\varphi = \varphi(E)$

$$UH_oU^{-1}\varphi(E) = E\varphi(E) \tag{3-10}$$

and, hence, (3-5) follows.

We shall now construct two possible examples of D which we call D_+ and D_-. They provide us with two examples for Φ, namely

$$\Phi_- = U^{-1}D_+ \qquad \text{and} \qquad \Phi_+ = U^{-1}D_- \tag{3-11}$$

Note the change in the labels + and - when going from the spaces $D_\pm$ to $\Phi^{\mp}$; the reason for this notation will become clear later.

We define D_+ as the linear space of functions satisfying the following conditions:

1. Any $\varphi(E) \in D_+$ is the restriction to R^+ of a function in H_+ (see Chapter II).
2. If $\varphi(E) \in D_+$, then $E^n\varphi(E) \in D_+$, n = 0,1,. . . Equivalently $\hat{E}D_+ \subset D_+$.
3. D_+ is dense in $L^2(R^+)$ in the sense of the Hilbert space topology on $L^2(R^+)$.
4. $\hat{E}$ is e.s.a. on D_+.
5. We endow D_+ with a complete nuclear (metrizable[4)]) topology such that $\hat{E}$ is continuous on D_+.
6. If $\varphi_n(E)$ is a sequence of functions in D_+ which converges to $\varphi(E) \in D_+$ with respect to the topology on D_+, then $\varphi_n(E) \to \varphi(E)$ with respect to the Hilbert space topology. In other words, we define the topology on D_+ to be finer (stronger) than the Hilbert space topology which D_+ possesses as a subspace of $L^2(R^+)$.

Why do we need these conditions? First of all, we see that they make the triplet

$$D_+ \subset L^2(R^+) \subset (D_+)^x \tag{3-12}$$

into a R.H.S. The reason for requiring condition 1 will be discussed later. The other conditions, 2 through 6, have the following consequences:

2 implies $H_0\Phi_- \subset \Phi_-$. In fact

$$\hat{E}D_+ \subset D_+ \Rightarrow U^{-1}\hat{E}D_+ \subset U^{-1}D_+ = \Phi_- \Rightarrow U^{-1}\hat{E}D_+ = (U^{-1}\hat{E}U)U^{-1}D_+$$

$$= H_0\Phi_- \subset \Phi_-. \tag{3-13}$$

Concerning 3, a unitary map transforms dense sets into dense sets. Thus, if D_+ is dense in $L^2(R^+)$, then $U^{-1}D_+ = \Phi_-$ is dense in $\mathcal{H}_0$.

4 implies that H_0 is e.s.a. on Φ_-.

5. If the topology on D_+ is metrizable[4] we can easily transport it to Φ_+ by means of U. A sequence $\{\varphi_n\} \subset \Phi_-$ converges to $\varphi \in \Phi_-$ if and only if $U\varphi_n \to U\varphi$ in D_+. In this way, Φ_- is endowed with a topology which has exactly the same properties as the topology of D_+. In particular Φ_- is a complete, metrizable, nuclear space.

6. If the topology on D_+ is finer (stronger) than the topology that D_+ has as a subspace of $L^2(R^+)$, then the same property holds for the topology on Φ_- with respect to the topology that Φ_- possesses as a subspace of $\mathcal{H}_0$.

Thus, as a consequence of conditions 2 through 6, it follows that

$$\Phi_- \subset \mathcal{H}_0 \subset (\Phi_-)^x \tag{3-14}$$

is a R.H.S. for which H_0 has the properties stated above on page 74.

We shall now construct the function space D_+ which has the above state properties 1 through 6. We shall do it in several steps, the first one will be the construction of another space Δ_+. We start with the space

$S(R^-) = \{$set of functions in S which are zero on $R^+\}$

$S(R^-)$ is a linear space and is dense in $L^2(R^-)$. $S(R^-)$ is a closed subspace of S and hence complete with respect to the topology of S.

By definition

$$\Delta_+ \equiv \mathcal{F}\ [S(R^-)] \tag{3-15}$$

The linear space Δ_+ is the set of the Fourier transforms of the functiosn in $S(R^-)$. We need a better characterization for Δ_+.

Note that $\mathcal{F}\ [S(R^-)] \subset S$ because $S(R^-) \subset S$ and the Fourier transform of a function in S is also in S. On the other hand, we know that according to Theorem II (Paley-Wiener):

$\mathcal{F}\ [S(R^-)] \subset \mathcal{H}_+$

Consequently:

$$\Delta_+ = \mathcal{F}\ [S(R^-)] \subset S \cap \mathcal{H}_+ \tag{3-16}$$

Pick now $f \in S \cap \mathcal{H}_+$. Its inverse Fourier transform has the following properties:

1. It belongs to $L^2(R^-)$ because of Theorem II (Paley-Wiener).

2. It is a function in S and, therefore, has to be in $S(R^-)$.

This yields

$$S \cap H_+ \subset F\,[S(R^-)] \tag{3-17}$$

Hence

$$\Delta_+ = S \cap H_+ \tag{3-18}$$

Δ_+ has the following properties:

1. From (3-18) follows obviously: $\Delta_+ \subset H_+$.
2. If $f(E) \in \Delta_+$, then also $E^n f(E) \in \Delta_+$; $n = 0,1,2,\ldots$

To prove it, let $f(E) \in \Delta_+$. Its inverse Fourier transform is $\check{f}(x) = F^{-1} f(E) \in S(R^-)$. If a function belongs to S, its derivatives are also in S. Thus,

$$(-i)^n \frac{d^n}{dx^n} \check{f}(x) \in S(R^-) \tag{3-19}$$

The Fourier transform of this derivative is $E^n f(E)$ which therefore belongs to Δ_+.

3. Δ_+ is dense in H_+. Proof: We have stated (without prof) that $S(R^-)$ is dense in $L^2(R^-)$. The Fourier transform is a unitary mapping from $L^2(R^-)$ into H_+ and, therefore, preserves denseness.

4. Δ_+ is a subspace of S. A sequence $\{f_n(E)\} \subset \Delta_+$ will converge to $f(E) \in \Delta_+$ iff $f_n(E) \rightarrow f(E)$ with respect to the topology in S. Δ_+ inherits the topology of S and is a closed subspace of S. Hence Δ_+ is complete. Since any subspace of a nuclear space is nuclear, Δ_+ is a complete metrizable nuclear space.

5. If $f_n(E) \rightarrow f(E)$ in S, it converges also in $L^2(R)$. We can see this from (1-110) with m = 0. There are more convergent sequences in $L^2(R)$ than there are in S. Thus, the topology on S is finer (stronger) than the topology on $L^2(R)$.

From these properties follows that

$$\Delta_+ \subset H_+ \subset (\Delta_+)^x \tag{3-20}$$

is a R.H.S.

Now, we can define D_+:

$D_+ = \{$Restrictions to R^+ of functions in $\Delta_+\}$

Thus, $\varphi(E) \in D_+$ if it is a complex function taking values on R^+. These values are those taken by some function $f(E) \in \Delta_+$ on R^+.

From the van Winter theorem, Theorem II, we know that any function in H_+ and therefore in particular any function in Δ_+, is completely determined by its values on R^+. Therefore, there exists a one to one mapping between Δ_+ and D_+. If we call θ this mapping, we can write

$$\theta[\Delta_+] = D_+ \tag{3-21}$$

With (3-21), we can define the convergence in $D_+ : \varphi_n(E) \to \varphi(E)$ in D_+ iff $f_n(E) \equiv \theta^{-1}\varphi_n(E) \to \theta^{-1}\varphi(E) \equiv f(E)$ in Δ_+. We recall that $f_n(E) \to f(E)$ in Δ_+ iff

$$\int_{-\infty}^{\infty} |N^m f_n(E) - N^m f(E)|^2 \, dE \to 0 \ ; \ n\to\infty \ ; \ m = 0,1,2,... \tag{3-22}$$

$$\text{where } N = ax^2 + b\frac{d^2}{dx^2} + c \ \text{ (see (1-111))} \tag{3-22}$$

Thus, $\varphi_n(E) \to \varphi(E)$ in D_+ iff $N^m f_n(E) \to N^m f(E)$ with respect to the Hilbert space norm for every $m = 0,1,2,\ldots$, where $\varphi_n(E) = \theta f_n(E)$ and $\varphi(E) = \theta f(E)$. This sense of convergence is rather peculiar and non intuitive, since after the definition of D_+ one would expect that $\varphi_n(E) \to \varphi(E)$ iff

$$\int_{0}^{\infty} |N^m \varphi_n(E) - N^m \varphi(E)|^2 \, dE \to 0 \, ; \, n\to \infty \, ; \, m = 0,1,2,... \tag{3-23}$$

This does not need to be true, because (3-22) $\Rightarrow$ (3-23), but the converse is in general not true.

To see it, note that

$$\int_{-\infty}^{\infty} |N^m(\varphi(E) - \varphi(E))|^2 \, dE = \int_{-\infty}^{0} |N^m(\varphi_n(E) - \varphi(E))|^2 \, dE + \int_{0}^{\infty} |N^m(\varphi_n(E) - \varphi(E))|^2 \, dE$$

If the left hand side of this equation is small, so is each term in the right hand side. However, if only one term in the right is small this does not imply that so is the other and hence the integral in the left.

One can see easily that D_+, as defined by (3-21) fulfills 1 through 6 on page 77.

1. comes from the definition of D_+.

2. If $f(E) \in \Delta_+$, $E^n f(E) \in \Delta_+$. Obviously, we have from the definition of θ

$$\theta[E^n f(E)] = E^n \theta[f(E)] \tag{3-24}$$

If $\psi(E) \in D_+$, then $\theta^{-1}[\psi(E)] \in \Delta_+ \Rightarrow E^n\theta^{-1}[\psi(E)] \in \Delta_+$ and

$$\theta[E^n\theta^{-1}\psi(E)] = E^n\theta[\theta^{-1}\psi(E)] = E^n\psi(E) \in D_+ \tag{3-25}$$

3. To prove that D_+ is dense in $L^2(R^+)$ with respect to the Hilbert space topology, we first define:

H_{++} = {Restrictions to R^+ of functions in H_+}.

Equivalently, $\varphi(E) \in H_{++}$ if and only if there exists a function $\varphi(z) \in H_+$ such that the complex numnbers given by $\varphi(E)$ are the values of $\varphi(z)$ for real non negative values of E. As a consequence of the van Winter theorem, there is a one to one and onto mapping between H_+ and H_{++}.

It has been proven that H_{++} is dense in $L^2(R^+)$ with respect to the Hilbert space topology.[5)]

To prove that D_+ is dense in $L^2(R^+)$, we recall that Δ_+ is dense in H_+. As a consequence, if $f(E) \in H_+$ there exists a sequence $\{f_n(E)\} \subset \Delta_+$ such that

$$\int_{-\infty}^{\infty} |f(E) - f_n(E)|^2 \, dE \to 0 \tag{3-26}$$

The integral in (3-26) is bigger or equal to

$$\int_{0}^{\infty} |f(E) - f_n(E)|^2 \, dE \to 0 \tag{3-27}$$

and therefore (3-27) goes also to zero whenever $n \to \infty$.

Now, let $\eta(E) \in H_{++}$. There exists a function f(E) in H_+ such that its values for (almost) every $E \in R^+$ coincide with those of $\eta(E)$. Since $f_n(E) \in \Delta_+ \ \forall\, n$, $\varphi_n(E) \equiv \theta[f_n(E)] \in D_+$ and

$$\int_{0}^{\infty} |\eta(E) - \varphi_n(E)|^2 \, dE = \int_{0}^{\infty} |f(E) - f_n(E)|^2 \, dE \to 0 \tag{3-28}$$

Thus, $\varphi_n(E) \to \eta(E)$ in $L^2(R^+)$. We have proven that for any $\eta(E) \in H_{++}$, there exists a sequence $\{\varphi_n(E)\} \subset D_+$ such that it converges to $\eta(E)$. Thus, D_+ is dense in H_{++} Now, we use the result mentioned earlier on the denseness of (H_{++}) in $L^2(R^+)$ and the properties of the Hilbert space topology to conclude that D_+ is dense in $L^2(R^+)$.

4. E is indeed e.s.a. on D_+.

The proof is based on the following result[6]:

Let A be a symmetric operator defined on a domain D dense in the Hilbert space H. Let A^+ be the adjoint of A. A is e.s.a. on D iff the equations

$$A^+f = \pm\, if \tag{3-29}$$

do not have solutions in H different from zero.

5. The topology on D_+ is complete, nuclear and metrizable. To see that E is continuous on D_+, consider a sequence in D_+, $\psi_n(E) \to \psi(E) \in D_+$ according to the topology on D_+. $E\psi_n(E) = E\psi_n(E) \in D_+$ and $\theta^{-1}[E\psi_n(E)] \in \Delta_+$. Thus

$$\theta^{-1}[E\psi_n(E)] = E\theta^{-1}[\psi_n(E)] \to E\theta^{-1}[\psi(E)] = \theta^{-1}[E\psi(E)] \tag{3-30}$$

where the limit should be taken with respect to the topology on Δ_+. Hence

$$E\psi_n(E) = E\psi_n(E) \to E\psi(E) = E\psi(E) \tag{3-31}$$

in D_+, which proves the continuity of E on D_+.

6. Let $\varphi_n(E) \to \varphi(E)$ in D_+. Then $\theta^{-1}\varphi_n(E) \to \theta^{-1}\varphi(E)$ in Δ_+, i.e., in S. So, $\theta^{-1}\varphi_n(E) \to \theta^{-1}\varphi(E)$ in H_+. This means

$$\int_{-\infty}^{\infty} |\theta^{-1}\varphi_n(E) - \theta^{-1}\varphi(E)|^2\, dE \to 0 \;;\; \int_0^{\infty} |\varphi_n(E) - \varphi(E)|^2 \to 0 \tag{3-32}$$

and, hence, $\varphi_n(E) \to \varphi(E)$ in $L^2(R^+)$, which proves 6 .

<u>The Second Solution.</u>

After we have constructed the space D_+ and, therewith, the triplets (3-12) and (3-14), we can easily define another space D_-, which is closely related to D_+. This new space D_- will provide us with a new RHS satisfying also requirements 1-3 listed at the beginning of the present chapter.

We define the space

$D_- = \{$complex conjugates of the functions in $D_+\}$.

The functions in D_+ are restrictions to R^+ of functions in $\Delta_+ = S \cap H_-$. The complex conjugate of the functions in H_+ are the functions in H_-. As a consequence, the functions in D_- are the restrictions to R^+ of the functions in $S \cap H_-$. If we define

$$\Delta_- = S \cap H_- \tag{3-33}$$

Then, by arguments similar to those used above, we can prove that

$$\Delta_- = F\ [S(R^+)] \tag{3-34}$$

where $S(R^+) = \{$set of functions in S which are zero on $R^-\}$.

All this suggests another definition of D_- which is the analogue of the definition of D_+ on p.

$D_- = \{$restrictions to R^+ of the functions in $\Delta_- = S \cap H_-\}$.

We endow D_- with a topology in exactly the same way as we have done for D_+. We first define the operation "restriction to R^+," denoted by θ. Then, $\{\varphi_n(E)\} \subset D_-$ converges to $\varphi(E) \in D_-$ if and only if $\theta^{-1}[\varphi_n(E)] \rightarrow \theta^{-1}[\varphi(E)]$ in Δ_-, i.e., in S (Δ_- is a closed subspace of S, as well as Δ_+). D_- fulfills conditions 2 through 6 on p 114-5 and 1 has to be replaced by:

1'. Every function in D_- is the restriction to R^+ of a function in H_-.

As a consequence of these conditions, we see that

$$D_- \subset L^2(R^+) \subset (D_-)^x \tag{3-35}$$

is a RHS.

If we now define as in (3-11)

$$\Phi_+ = U^{-1}D_-$$

then Φ_+ satisfy the same premises as Φ_-. The triplet

$$\Phi_+ \subset H_0 \subset (\Phi_+)^x \tag{3-36}$$

is then another example of a RHS which fulfills the requirements formulated at the beginning of the present chapter. As intermediate steps, we have therewith also found two more RHS of functions.

$$D_- \subset L^2(R^+) \subset (D_-)^x \tag{3-37}$$

and

$$\Delta_- \subset H_- \subset (\Delta_-)^x \tag{3-38}$$

(3-37) is a realization of (3-36). However, (3-38) is not a realization of (3-36) because (3-36) and (3-38) are not equivalent RHS. The reason for this is that

$$\Phi_+ = U^{-1}\theta\Delta_- \tag{3-39}$$

but θ is not unitary. Hence, $U^{-1}\theta$ is <u>not</u> a unitary transformation. In addition, $U^{-1}\theta$ cannot be uniquely extended to be a mapping from H_- into $L^2(R^+)$. For the same reason (3-12) is a realization of (3-14) whereas (3-20) is not.

Finally, the following relations can be proven

$$\Delta_+ \cap \Delta_- = \{0\} \tag{3-40}$$

$$D_- \cap D_+ \neq \{0\}\ ; \qquad \Phi_- \cap \Phi_+ \neq \{0\} \tag{3-41}$$

(3-40) says that the zero function is the only one which is simultaneously an element of Δ_+ and of Δ_-. (3-40) follows from

$$H_+ \cap H_- = \{0\} \tag{3-40A}$$

A proof of (3-41) can be found in Reference 7.

We want to close this section with a remark concerning our notation. The reason for the signs plus and minus in $D_\pm$ and $\Delta_\pm$ has now become obvious. These signs have been inherited from the signs in $H_\pm$, which in turn come from $C^\pm$. The sign changes in definitions (3-11) are then needed in order to use for Φ_- and Φ_+ a notation inherited from the Moller wave operators in the standard notation of scattering theory (cf. eqs. (3-42) in the next section).

II. THE SPACES Φ^+ AND Φ^-

In all that follows, we assume asymptotic completeness, i.e., $H_{ac} = \Omega^{\pm} H_o$ (see Chapter II).

Call

$$\Phi^+ = \Omega^+\Phi_+ ; \qquad \Phi^- = \Omega^-\Phi_- \tag{3-42}$$

We topologize Φ^+ and Φ^- as follows: A sequence of vectors (φ_n) converges to φ in Φ^+ iff $(\Omega^+)^{-1}\varphi_n \to (\Omega^-)^{-1}\varphi$ with respect to the topology on Φ_+.

Analogously, $\psi_n \to \psi$ in Φ^- iff $(\Omega^-)^{-1}\psi_n \to (\Omega^-)^{-1}\psi$ in Φ_-.

These topologies are good enough so that

$$\Phi^+ \subset H_{ac} \subset (\Phi^+)^x \text{ and } \Phi^- \subset H_{ac} \subset (\Phi^-)^x \tag{3-43}$$

become RHS. In fact, since $\Phi_\pm \subset H_o$, it follows that $\Omega^\pm\Phi_\pm \subset \Omega^\pm H_o$ or equivalently $\Phi^\pm \subset H_{ac}$. Both $\Phi_\pm$ are dense in H_o and $\Omega^\pm$ are unitary mappings from H_o onto H_{ac}. Consequently, $\Phi^\pm$ are dense in H_{ac}. $\Phi^\pm$ are complete, metrizable, nuclear spaces and their topology defined above finer than the topology they have as subspace of H_{ac}.

We shall define the extension of $\Omega^\pm$ to $(\Phi_\pm)^x$ in the following way:

$$\langle\Omega^+\varphi_+ | (\Omega^+)^x F_+\rangle = \langle\varphi_+ | F_+\rangle \qquad \forall\varphi \in \Phi_+,\ F_+ \in (\Phi_+) \tag{3-44-A}$$

$$\langle\Omega^-\varphi_- | (\Omega^-)^x F_-\rangle = \langle\varphi_- | F_-\rangle \qquad \forall\varphi \in \Phi_-,\ F_- \in (\Phi_-)^x . \tag{3-44-B}$$

After this definition, one can show that

$$(\Omega^+)^x F^+ \in (\Omega^+)^x \quad \text{and} \quad (\Omega^-)^x F_- \in (\Phi^-)^x . \tag{3-45}$$

To prove it, let $(\varphi_n)_-$ be a sequence in Φ_- converging to $\varphi_- \in \Phi_-$. Thus, $\Omega^-(\varphi_n)_-$ converges to $\Omega^-\varphi_-$ in Φ^-. Therefore

$$\langle\Omega^-(\varphi_n)_- | (\Omega^-)^x F_-\rangle = \langle(\varphi_n)_- | F_-\rangle \to \langle\varphi_- | F_-\rangle = \langle\Omega^-\varphi_- | (\Omega^-)^x F_-\rangle$$

or

$$\langle\Omega^-(\varphi_n)_- | (\Omega^-)^x F_-\rangle \to \langle\Omega^-\varphi_- | (\Omega^-)^x F_-\rangle .$$

Since $(\Omega^-)^x F_-$ is obviously linear, the desired result follows.

Furthermore, the mappings $(\Omega^\pm)^x$ are bijective (one to one and onto) and bicontinuous $\Omega^\pm$) and $(\Omega^\pm)^{-1}$ are continuous) so that

$$(\Omega^+)^x(\Phi_+)^x = (\Phi^+)^x \text{ and } (\Omega^-)^x(\Phi_-)^x = (\Phi^-)^x \tag{3-46}$$

and the continuity should be understood in the sense of the topology on the duals. We do not give the proof of this statement here.[8)]

In (2-11), we have defined H_l as the restriction of H to $\mathcal{H}_{ac}$. If we do so, we find

$$H_l\Phi^+ \subset \Phi^+ \quad \text{and} \quad H_l\Phi^- \subset \Phi^- . \tag{3-47}$$

The proof is the following:

$$H_l\Phi^\pm = H_l\Omega^\pm\Phi_\pm = \Omega^\pm H_o\Phi_\pm \subset \Omega^+\Phi_\pm = \Phi^\pm . \tag{3.48}$$

The second equality is (2-22). The other steps are obvious.

We shall now show that H_l is continuous on $\Phi^\pm$. Assume $\varphi_n \to \varphi$ in Φ_+. H_o is continuous on Φ_+, then $H_o(\Omega^+)^{-1}\varphi_n \to H_o(\Omega^+)^{-1}\varphi$. (2-22) gives $H_o(\Omega^+)^{-1} = (\Omega^+)^{-1}H_l$. Thus

$$(\Omega^+)^{-1}H_l\varphi_n \to (\Omega^+)^{-1}H_l\varphi \tag{3-49}$$

in Φ_+ and therefore

$$H_l\varphi_n \to H_l\varphi \tag{3-50}$$

in Φ^+. By the same arguments, using Ω^- instead of Ω^+ one proves that H_l is continuous on Φ^-.

H_l is e.s.a. on $\mathcal{H}_{ac}$ because unitary mappings preserve this property (see (2-22)).

Finally, H_l can be extended to $(\Phi^\pm)^x$ by

$$\langle H_l\varphi|F\rangle = \langle\varphi|(H_l)^x F\rangle \qquad \text{for } \varphi \in \Phi^+ \text{ or } \Phi^- \tag{3-51}$$

and $F \in (\Phi^+)^x$ or $(\Phi^-)^x$ respectively.

$(H_l)^x$ is continuous on $(\Phi^+)^x$ and on $(\Phi^-)^x$.[8)]

III. FUNCTIONALS FOR H_0 AND H_1

We already know that $H_0\Phi_+ \subset \Phi_+$ and that H_0 is continuous on Φ_+. Therefore, H_0 satisfies the conditions of the Gelfand-Maurin theorem. This implies the existence of a complete set of generalized eigenvectors, $|E_+\rangle \in (\Phi_+)^x$, of H_0:

$$(H_0)^x|E_+\rangle = E|E_+\rangle \qquad \forall E \in R^+ \tag{3-52-A}$$

$$(\varphi_+,\psi_+) = \int^\infty_0 \langle\varphi_+|E_+\rangle\langle E_+|\psi_+\rangle\, dE \qquad \forall \varphi_+, \psi_+ \in \Phi_+ \tag{3-53-A}$$

$$\langle\psi_+|E_+\rangle^* = \langle E_+|\psi_+\rangle \equiv \psi_-(E) \in D_- . \tag{3-54-A}$$

As H_0 is also continuous on Φ_-, there exists another set of generalized eigenvectors, $|E_-\rangle \in (\Phi_-)^x$, of H_0:

$$(H_0)^x|E_-\rangle \qquad \forall\, E \in R^+ \tag{3-52-B}$$

$$(\varphi_-,\psi_-) = \int^\infty_0 \langle\varphi_-|E_-\rangle\langle E_-|\psi_-\rangle\, dE \qquad \forall \varphi_-, \psi_- \in \Phi_- \tag{3-53-B}$$

$$\langle\psi_-|E_-\rangle^* = \langle E_-|\psi_-\rangle \equiv \psi_+(E) \in D_+ . \tag{3-54-B}$$

On the other hand, since $|E_\pm\rangle = (\Phi_\pm)^x$, we can define

$$|E^+\rangle = (\Omega^+)^x\, |E_+\rangle \in (\Phi^+)^x; \quad |E^-\rangle = (\Omega^-)^x\, |E_-\rangle \in (\Phi^-)^x . \tag{3-55}$$

The importance of the vectors (3-55) comes from the next proposition.

<u>Proposition:</u>

$$(H_1)^x\, |E^\pm\rangle = E|E^\pm\rangle . \tag{3-56}$$

In addition, the vectors $|E^\pm\rangle$ when E runs through R^+, form a complete set of generalized eigenvectors for H_1 in $(\Phi^\pm)^x$.

<u>Proof</u>: First, we need to show that

$$(H_0)^x = [(\Omega^\pm)^x]^{-1}\, (H_1)^x(\Omega^\pm)^x \tag{3-57}$$

which are the extension of relations (2-22) to $(\Phi_\pm)^x$. Take $\varphi_\pm \in \Phi_\pm$ and $F_\pm \in (\Phi_\pm)^x$, then

$$\begin{aligned}\langle\varphi_\pm|(H_0)^x\, F_\pm\rangle &= \langle H_0\varphi_\pm|F_\pm\rangle \\ &= \langle(\Omega^\pm)^{-1}\, H_1\Omega^\pm\varphi_\pm\, |F_\pm\rangle \\ &= \langle H_1\Omega^\pm\varphi_\pm|(\Omega^\pm)^x\, F_\pm\rangle = \langle\Omega^\pm\varphi_\pm|(H_1)^x(\Omega^\pm)^x\, F_\pm\rangle \\ &= \langle\varphi_\pm|[(\Omega^\pm)^x]^{-1}\, (H_1)^x(\Omega^\pm)^x\, F_\pm\rangle .\end{aligned} \tag{3-58}$$

To show the validity of step three in the above equality, take for instance $\varphi^+ \in \Phi^+$ and $F_+ \in (\Phi_+)^x$. Then,

$$\langle(\Omega^+)^{-1}\, \varphi^+|F_+\rangle = \langle\Omega^+(\Omega^+)^{-1}\, \varphi^+|(\Omega^+)^x\, F_+\rangle = \langle\varphi^+|(\Omega^+)^x\, F_+\rangle . \tag{3-59}$$

To show the validity of step five, take

$F^+ \in (\Phi^+)^x$ and $\varphi^+ \in \Phi^+$

$\varphi_+ = (\Omega^+)^{-1}\,\varphi^+ \in \Phi_+$ and

$F_+ = [(\Omega^+)^x]^{-1}\,F^+ \in (\Phi_+)^x$.

The equation (3-44-A) also reads

$$\langle \varphi^+ | F^+ \rangle = \langle (\Omega +)^{-1}\,\varphi^+ | [(\Omega^+)^x]^{-1}\,F^+ \qquad (3\text{-}44\text{-}C)$$

Then, $(\Omega^+)^{-1}$

$$\langle \Omega^+\varphi_+ | F^+ \rangle = \langle (\Omega^+)^{-1}\,\Omega^+\varphi_+ | [(\Omega^+)^x]^{-1}\,F^+ \rangle$$

$$= \langle \varphi_+ | [(\Omega^+)^x]^{-1}\,F^+ \rangle \qquad (3\text{-}60)$$

whence step five follows.

Now, we can prove the proposition. It is obvious that (3-57) implies

$$(H_l)^x = (\Omega^\pm)^x\,(H_o)^x\,[(\Omega^\pm)^x]^{-1} \qquad (3\text{-}61)$$

Hence,

$$(H_l)^x\,|E^\pm\rangle = (\Omega^\pm)^x\,(H_o)^x\,[(\Omega^\pm)^x]^{-1}\,|E^\pm\rangle$$

$$= (\Omega^\pm)^x\,(H_o)^x\,|E_\pm\rangle = (\Omega^\pm)^x\,E|E_\pm\rangle$$

$$= E(\Omega^\pm)^x\,(H_o)^x\,|E_\pm\rangle = E|E^\pm\rangle\,. \qquad (3\text{-}62)$$

This proves the first part of the proposition. For the second, take $\varphi^\pm = \Omega^\pm\varphi_\pm$

$$(\varphi^\pm,\psi^\pm) = (\Omega^\pm\varphi_\pm,\Omega^\pm\psi_\pm) = (\varphi_\pm,\psi_\pm)$$

$$= \int^\infty_0 \langle \varphi_\pm | E_\pm \rangle \langle E_\pm | \psi_\pm \rangle\, dE$$

$$= \int^\infty_0 \langle \Omega^\pm\varphi_\pm | \Omega^\pm E_\pm \rangle \langle \Omega^\pm E_\pm | \Omega^\pm\psi_\pm \rangle\, dE$$

$$= \int^\infty_0 \langle \varphi^\pm | E^\pm \rangle \langle E^\pm | \psi^\pm \rangle\, dE \qquad (3\text{-}63)$$

which proves the completeness of $|E^\pm\rangle$ in $\Phi^\pm$ when E runs through R^+.

There are more generalized eigenvalues of H_o and H_l, than the $E \in R^+$ of (3-56). These generalized eigenvalues depend upon the definition of $\Phi_\pm$ and $\Phi^\pm$. They are very important for our final objective. Finding these generalized eigenvalues is our next goal.

Let ω be a complex number such that $\mathrm{Im}\,\omega \le 0$. Then, there exists a unique $|\omega_-\rangle \in (\Phi^-)^x$ such that

$$(H_o)^x\,|\omega_-\rangle = \omega|\omega_-\rangle\,. \qquad (3\text{-}64\text{-}A)$$

If $\mathrm{Im}\,\omega \ge 0$, there exists a unique $|\omega_+\rangle \in (\Phi_+)^x$ such that

$$(H_o)^x\,|\omega_+\rangle = \omega|\omega_+\rangle\,. \qquad (3\text{-}64\text{-}B)$$

Before we can find these generalized eigenvectors, we need to extend the mapping θ to the duals.

Take e.g. $\varphi_- \in \Phi_-$. $U\varphi_- = \varphi_+(E) \in D_+$ and $\theta^{-1}\varphi_+(E) = g_+(E) \in \Delta_+$. We can extend θ to become a bijective and bicontinuous mapping, θ^x, from $(\Delta_+)^x$ onto $(D_+)^x$. This can be done through the defining relation:

$$\langle g_+(E)|h_+\rangle = \langle \theta g_+(E)|\theta^x h_+\rangle$$
$$\forall g_+(E) \in \Delta_+ ; \qquad \forall h_+ \in (\Delta_+)^x . \qquad (3\text{-}65\text{-}A)$$

It is an easy exercise to prove that θ^x is indeed an extension of θ. Note that (3-65-A) implies that

$$\langle \theta^{-1}\varphi_+(E)|(\theta^x)^{-1} f_+\rangle = \langle \varphi_+(E)|f_+\rangle$$
$$\forall \varphi_+(E) \in D_+ ; \qquad \forall f_+ \in (D_+)^x . \qquad (3\text{-}66)$$

Analogously, when we consider θ as a bijection from Δ_- onto D_-, we can define θ^x as

$$\langle g_-(E)|h_-\rangle = \langle \theta g_-(E)|\theta^x h_-\rangle \qquad (3\text{-}65\text{-}B)$$

which is a bicontinuous bijection from $(\Delta_-)^x$ onto $(D_-)^x$. Formula (3-63) is also valid here, with a change from the + sign to the - sign.

We are going now to find the vectors $|\omega_\pm\rangle$ in (3-64) for the case in which Im $\omega \neq 0$. For any $\omega \in C^-$ with Im $\omega < 0$, let $F_{\omega-}$ be a functional such that

$$\langle \varphi_-|F_{\omega-}\rangle = [g_+(\omega)]^* \qquad \forall \varphi_- \in \Phi_- , \qquad (3\text{-}67\text{-}A)$$

where $g_+(E) = \theta^{-1} U \varphi_- = \theta^{-1}\varphi_+(E)$ (see diagram) is a function in Δ_+ and, therefore, $g_+(z)$ is analytic when Im $z > 0$. Moreover, $g_+(E) \in H_+$ and $[g_+(E)]^* \in H_-$ (see Chapter II, Section III), and consequently $[g_+(\omega)]^*$ is well defined as the value of the analytic function $[g_+(z)]^*$ at the point $\omega \in C^- - R$. By using the Titchmarsh theorem, one can prove that $F_{\omega-}$ belongs to $(\Phi_-)^x$, i.e., it is a continuous antilinear functional on Φ_-.

The proof goes as follows: Let $\varphi_n \to \varphi$ in Φ_-. We have to prove that $\langle \varphi_n - \varphi|F_{\omega-}\rangle \to 0$ or equivalently $\langle \varphi_n|F_{\omega-}\rangle \to \langle \varphi|F_{\omega-}\rangle$. If we call

$g_n(E) = \theta^{-1} U\varphi_n$; $g(E) = \theta^{-1} U\varphi$.

Then, $g(E), g_n(E) \in \Delta_+$ and

$$\langle \varphi_n - \varphi|F_{\omega-}\rangle = [g_n(\omega)]^* - [g(\omega)]^*$$
$$= - (2\pi i)^{-1} \int_{-\infty}^{\infty} \{[g_n(E)]^* - [g(E)]^*\} (E - \omega)^{-1} dE . \qquad (3\text{-}68)$$

We made use of the Titchmarsh theorem in the last step of (3-68).

Since Im $\omega \neq 0$, $(E - \omega)^{-1}$ belongs to $L^2(R)$. In $L^2(R)$, we have the following scalar product:

$$(g_n(E) - g(E), (E - \omega)^{-1}) = \int_{-\infty}^{\infty} \{[g_n(E)]^* - [g(E)]^*\}(E - \omega)^{-1} dE . \qquad (3\text{-}69)$$

Then, from (3-68) and (3-69) follows

$$|\langle \varphi_n - \varphi|F_{\omega-}\rangle| = |(g_n(E) - g(E), (E - \omega)^{-1})| (2\pi)^{-1}$$

and, using the Cauchy-Schwarz inequality, we obtain:

$$|\langle \varphi_n - \varphi|F_{\omega-}\rangle| \leq \|g_n(E) - g(E)\| \, \|(E - \omega)^{-1}\| (2\pi)^{-1} . \qquad (3\text{-}70)$$

Since θ^{-1} and U are continuous, it follows that, if $\varphi_n \to \varphi$ in Φ_-, then $g_n(E) \to g(E)$ in Δ_+. Consequently, from $\varphi_n \to \varphi$ follows $\|g_n(E) - g(E)\| \to 0$ and hence by (3-66):

$\langle\varphi_n - \varphi|F_{\omega-}\rangle \to 0$. This proves the continuity of $F_{\omega-}$ in Φ_-.

For any $\omega \in C^+$ with Im $\omega > 0$, let $F_{\omega+}$ be a functional such that

$$\langle\varphi_+|F_{\omega+}\rangle = [g_-(\omega)]^* \qquad \forall\varphi_+ \in \Phi_+ \tag{3-67-B}$$

where $g_-(E) = \theta^{-1} U\varphi_+$ is a function in Δ_-. Therefore, by similar argument as above, $[g_-(\omega)]^*$ is a well defined analytic function if Im $\omega > 0$ and $F_{\omega+} \in (\Phi_+)^x$.

If $\omega \in R^+$, then $F_{\omega\pm} = |E_\pm\rangle$. If $\omega \in R^+$, we can still define an $F_{\omega\pm}$ as in (3-67) which belongs to $(\Phi_\pm)^x$. The proof is, however, different.

Let us now compute $(H_o)^x F_{\omega-}$. For this purpose we calculate

$$\langle\varphi_-|(H_o)^x F_{\omega-}\rangle = \langle H_o\varphi_-|F_{\omega-}\rangle = [f_+(\omega)]^* \tag{3-71}$$

where $f_+(E) = \theta^{-1} U(H_o\varphi_-)$, which belongs to Δ_+, as $H_o\varphi_- \in \Phi_-$ (see next diagram). Since $U\varphi_- = \varphi_+(E) \in D_+$, we can write

$$\begin{aligned} f_+(E) &= \theta^{-1} UH_oU^{-1} U\varphi_- = \theta^{-1}E\, U\varphi_- = \theta^{-1}E\varphi_+(E) \\ &= \theta^{-1}E\varphi_+(E) = E\theta^{-1}\varphi_+(E) = Eg_+(E) \end{aligned} \tag{3-72}$$

where $g_+(E) = \theta^{-1}\varphi_+(E) = \theta^{-1} U\varphi_-$. In the second identity of (3-72), equation (3-2) was used. The value of $[f_+(\omega)]^*$ can be obtained by using the Titchmarsh theorem and the fact that $[f_+(E)]^* \in H_-$:

$$\begin{aligned} [f_+(\omega)]^* &= -(2\pi i)^{-1} \int_{-\infty}^{\infty} [f_+(E)]^*(E - \omega)^{-1}\, dE \\ &= -(2\pi i)^{-1} \int_{-\infty}^{\infty} E[g_+(E)]^*(E - \omega)^{-1}\, dE = \omega[g_+(\omega)]^* . \end{aligned} \tag{3-73}$$

The last step in (3-73) holds because $E[g_+(E)]^* \in H_-$. This, using (3-67-A) and (3-71), we obtain from (3-69)

$$\langle\varphi_-|(H_o)^x F_{\omega-}\rangle = \omega[g_+(\omega)]^* = \omega\langle\varphi_-|F_{\omega-}\rangle . \tag{3-74}$$

Since $\varphi_- \in \Phi_-$ is arbitrary, (3-74) can be written

$$(H_o)^x F_{\omega-} = \omega F_{\omega-} . \tag{3-75}$$

Furthermore, the generalized eigenvalues of H_o are degeneracy free. Therefore, $F_{\omega-}$ will differ from $|\omega_-\rangle$ of (3-64-A) by at most a phase and we can identify

$$F_{\omega-} = |\omega_-\rangle \tag{3-76}$$

(3-75) then proves (3-64-A). The existence and uniqueness of $|\omega_+\rangle$ in (3-64-B) can be proven analogously. In Chapter IV, Section III we shall discuss the case when ω is real and negative.

Now we define:

$$|\omega^{\pm}\rangle = (\Omega^{\pm})^{x}\, |\omega_{\pm}\rangle\,. \tag{3-77}$$

One can easily prove that

$$(H_1)^{x}\, |\omega^{\pm}\rangle = \omega|\omega^{\pm}\rangle\,. \tag{3-78}$$

In the diagram, we display the spaces used so far and summarizes the relations among them.

<u>Remarks on the diagram</u>: We define U^x on $F_{\pm} \in (\Phi_{\pm})^x$ by means of the following formula

$$\langle U\varphi_{\pm}|U^{x}\, F_{\pm}\rangle = \langle \varphi_{\pm}|F_{\pm}\rangle \tag{3-79}$$

and U^x is a bicontinuous bijection from $(\Phi_{\pm})^x$ onto $(D\ \ )^x$.

On the other hand, the mapping "restriction to R^{+}" θ, does not relate $H_{\pm}$ to $L^2(R^+)$. In fact, we can find functions in $L^2(R^+)$ which are not a restriction of a Hardy class function as defined on Chapter II. However, if $\varphi(E) \in L^2(R^+)$, it also belongs to $(D_+)^x$ (and to $(D_-)^x$) and, therefore, $[\theta^x]^{-1}\varphi(E) \in (\Delta_+)^x$. $[\theta^x]^{-1}\varphi(E)$ does not always belong to H_+ or even to $L^2(R)$. Conversely, if $g_{\pm}(E) \in H_{\pm} \Rightarrow g_{\pm}(E) \in (\Delta_{\pm})^x$ and hence $\theta^x g_{\pm}(E) \in (D_{\pm})^x$, which does not mean that $\theta^x g_{\pm}(E)$ belongs to $L^2(R^+)$. This apparently peculiar behavior of θ^x comes from the fact that θ is <u>not</u> a unitary mapping from $\Delta_{\pm}$ onto $D_{\pm}$.

In the following, we shall derive an equation equivalent to (3-63) and (3-67) valid in the triplets (3-20) and (3-38). The importance of this result will become clear at the end of the next chapter. We start with (3-77) from which follows:

$$\langle \varphi^{\pm}\,|\,\omega^{\pm}\rangle = \langle \varphi_{\pm}\,|\,\omega_{\pm}\rangle = [g\ \ (\omega)]^{*} \tag{3-80}$$

$$\varphi_{\pm} \in \Phi_{\pm}\,;\quad \varphi^{\pm} = \Omega^{\pm}\varphi_{\pm}\,;\ \omega \in C^{\pm}\,.$$

Now, take $\operatorname{Im}\omega < 0$, use the fact that $[g_+(E)]^{*} \in H_-$ and apply the Titchmarsh theorem (2-31B) then

$$[g_+(\omega)]^{*} = -\,(2\pi i\,)^{-1} \int_{-\infty}^{\infty} [g_+(E\,)\,]^{*}(E - \omega\,)^{-1}\, dE\ . \tag{3-81}$$

The function $-(2\pi i\,)^{-1}\ (E - \omega)^{-1}$ is square integrable. In other words, it belongs to $L^2(R)$ and so does $g_+(E)$. (3-81) is thus the scalar product on $L^2(R)$ of these two functions:

$$(3\text{-}81) = (g_+(E)\,,\ -(2\pi i)^{-1}(E - \omega)^{-1})\,. \tag{3-82}$$

Since $g_+(E) \in \Delta_+$ and $-(2\pi i)^{-1}(E - \omega)^{-1} \in (\Delta_+)^x$ this can also be written

$$(3\text{-}82) = \langle g_+(E) | -(2\pi i)^{-1}(E - \omega)^{-1}\rangle = [g_+(\omega)]^*\,. \tag{3-83}$$

(3-80) reminds us of (3-67) with

$$\varphi = U^{-1}\theta[g_\pm(E)]\,. \tag{3-84}$$

Combining (3-80) and (3-83) yields

$$\langle \varphi^- | \omega^- \rangle = \langle g_+(E) | -(2\pi i)^{-1}(E - \omega)^{-1}\rangle\,. \tag{3-85-A}$$

With the help of (3-84) and $\varphi^\pm = \Omega^\pm\, \varphi^\pm$ one thus obtains

$$|\omega^-\rangle = (\Omega^-)^x (U^{-1})^x\, \theta^x\{-(2\pi i)^{-1}(E - \omega)^{-1}\}\,. \tag{3-86-A}$$

Similar arguments for $\mathrm{Im}\,\omega < 0$ give the corresponding results:

$$\langle \varphi^+ | \omega^+ \rangle = \langle g_-(E) | (2\pi i)^{-1}(E - \omega)^{-1}\rangle = [g_-(\omega)]^* \tag{3-85-B}$$

and

$$|\omega^+\rangle = (\Omega^+)^x (U^{-1})^x\, \theta^x\{(2\pi i)^{-1}(E - \omega)^{-1}\}\,. \tag{3-86-B}$$

We shall use the following notation

$$G_+(\omega) = -(2\pi i)^{-1}(E - \omega)^{-1}\,;\ \mathrm{Im}\,\omega < 0 \tag{3-87-A}$$

$$G_-(\omega) = (2\pi i)^{-1}(E - \omega)^{-1}\,;\ \mathrm{Im}\,\omega > 0\,. \tag{3-87-B}$$

We know already that $G_\pm(\omega) \in (\Delta_\pm)^x\ \forall\omega \in \{C^\mp - R\}$. Furthermore they are square integrable functions. In fact, one can see using the definition of the Hardy class functions that

$$G_\pm(\omega) \in \mathcal{H}_\pm\,;\quad \omega \in C^\mp\,. \tag{3-88}$$

Note, however, that in spite of (3-88):

$$\theta^x\, G_\pm(E) \tag{3-89}$$

do in general not belong to $L^2(R)$.

Now, we define

$$\hat{\varepsilon} = \theta^{-1} E\theta \tag{3-90}$$

and by using (2-22) and $E = UH_0U^{-1}$ we obtain

$$\hat{\varepsilon} = \theta^{-1}\, U(\Omega^\pm)^{-1} H_1\, \Omega^\pm\, U^{-1}\theta\,. \tag{3-91}$$

Pick a $g_\pm(E) \in \Delta_\pm$, then $\theta g_\pm(E) \equiv \varphi_\pm(E) \in D_\pm$ and we have

$$\begin{aligned}\hat{\varepsilon} g_\pm(E) &= \theta^{-1}\hat{E}\theta g_\pm(E) = \theta^{-1}\hat{E}\varphi_\pm(E) = \\ &= \theta^{-1}E\varphi_\pm(E) = E\theta^{-1}\varphi_\pm(E) = Eg_\pm(E)\end{aligned} \tag{3-92}$$

or

$$\hat{\varepsilon} g_\pm(E) = Eg_\pm(E)\,,\quad E \in R\,. \tag{3-93}$$

The next formula extends $\hat{\varepsilon}$ to $(\Delta_\pm)^x$. Let $G_\pm \in (\Delta_\pm)^x$

$$\langle \hat{\varepsilon} g_\pm(E) \,|\, G_\pm \rangle = \langle g_\pm(E) \,|\, \hat{\varepsilon}^x G_\pm \rangle \qquad \forall g_\pm(E) \in \Delta_\pm \,. \tag{3-94}$$

Since $\hat{E}$ is continuous and θ bicontinuous, $\hat{\varepsilon}$ is continuous on $\Delta_\pm$. Also $\hat{\varepsilon}\Delta_\pm = \theta^{-1}\hat{E}\theta\Delta_\pm = \theta^{-1}ED_\pm \subset \theta^{-1}D_\pm = \Delta_\pm$. Therefore, $\hat{\varepsilon}^x$ in (3-94) is a well defined continuous operator on $(\Delta_\pm)^x$.

<u>Proposition</u>.

$$\hat{\varepsilon}^x G_\pm(\omega) = \omega G_\pm(\omega) \,. \tag{3-95}$$

<u>Proof</u>. Take, for instance, $g_+(E) \in \Delta_+$, $\text{Im}\,\omega < 0$.

$\langle \hat{\varepsilon} g_+(E) \,|\, G_+(\omega) \rangle = \langle E g_+(E) \,|\, G_+(\omega) \rangle =$

$= (E g_+(E), G_+(\omega)) = -(2\pi i)^{-1} \int_{-\infty}^{\infty} E[g(E)]^* (E - \omega)^{-1} \, dE =$

$$= \omega [g_+(\omega)]^* = \omega \langle g_+(E) \,|\, G_+(\omega) \rangle \,. \tag{3-96}$$

The third identity in (3-96) is true because $E[g_+(E)]^* \in H_-$ (see (2-31-B)). The fourth identity follows from (3-81). The proof for G_- is similar. This proposition will be used in the next chapter.

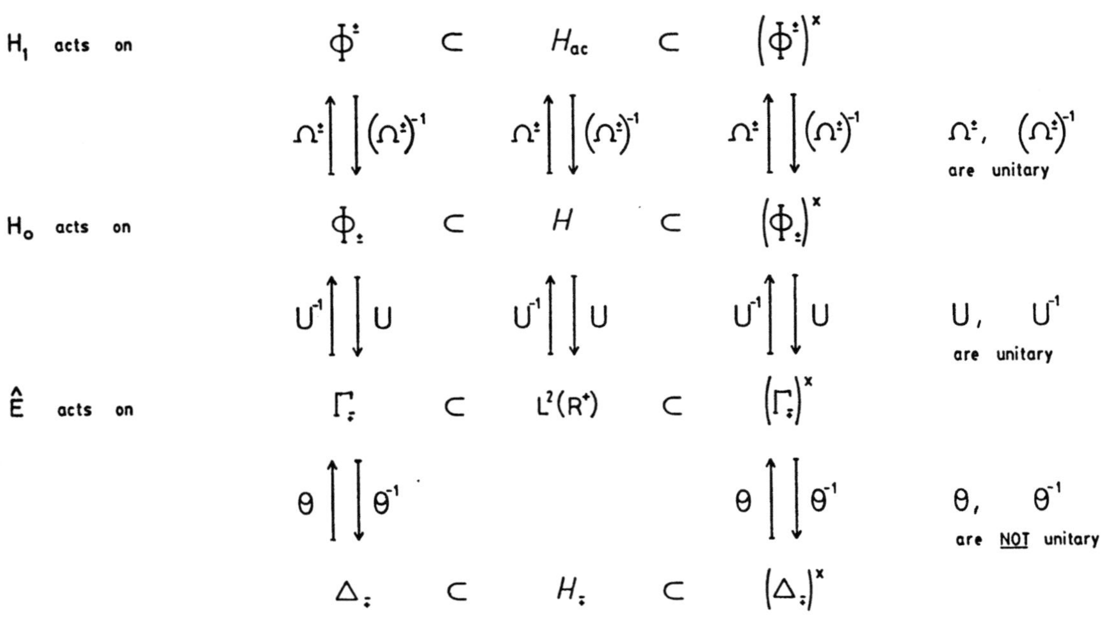

References for Chapter III

1.- See Appendix III for the definition of essential self-adjointness.

2.- See Reference 20 in Chapter II.

3.- As a result of the spectral theorem. See Reference 29 in Chapter I and Reference 22 in Chapter II.

4.- A linear topological space X is called metrizable if we can define a distance between the elements of X such that the topology that this distance endows to X coincides with that one X has as linear topological space. In particular, if the topology on X is defined by a countable number of norms, then X is metrizable.

5.- See Reference 17 in Chapter II

6.- See Reference 20 in Chapter II.

7.- M. Gadella, J. Math. Phys., 24 ,1462, (1983).

8.- M. Gadella, J. Math. Phys., 25 ,2481 (1984).

CHAPTER IV

I. THE RHS MODEL FOR DECAYING STATES

The objective of the present chapter is to describe our RHS model for resonances and virtual states. This chapter is divided in three sections. In the first one, we obtain the so called Gamow vectors for $l = 0$ [1].

To begin with, we here list sufficient conditions under which the construction of this model is possible. From now on, we consider a resonance scattering with the following properties:

1. The scattering is produced by a potential V. The total hamiltonian is given by $H = H_0 + V$ where H_0 is the hamiltonian for the free particle (we call H_0 the free hamiltonian). These operators act on a Hilbert space of states, $\mathcal{H}$.

2. H can have discrete eigenvalues and an absolutely continuous spectrum, but it cannot have a singular spectrum.[2] The continuous spectrum of H is equal to R^+. Then, the Hilbert space of states can be decomposed into the Hilbert space $\mathcal{H}_p$ spanned by the eigenvectors with discrete eigenvalues and its orthogonal complement $\mathcal{H}_{ac}$:

$$\mathcal{H} = \mathcal{H}_p \oplus \mathcal{H}_{ac} \tag{4-1}$$

We also recall that if $f \in \mathcal{H}_{ac}$ is in the domain of H then $Hf \in \mathcal{H}_{ac}$ and, therefore, $\mathcal{H}_{ac}$ is invariant under the action of H (although not every vector in $\mathcal{H}_{ac}$ may be in the domain of H). We call H_l the restriction of H to $\mathcal{H}_{ac}$.

3. The Moller wave operators (2-17) and (2-20) exist and asymptotical completeness holds. The S operator is defined in terms of the Moller wave operators by

$$S = (\Omega^-)^{-1}\,\Omega^+ . \tag{4-2}$$

Since we have asymptotic completeness, S is a unitary operator on $\mathcal{H}$ (see Chapter II, Section II and references therein).

4. We ignore possible additional quantum numbers and assume angular momentum conservation (and restrict ourselves to spinless projectiles). Then we can restrict ourselves to a fixed value of angular momentum . Note that in general

$$\mathcal{H} = \bigoplus_{l=0} \mathcal{H}_l \tag{4-3}$$

where $\mathcal{H}_l$ is the space of states with a fixed angular momentum. The angular momentum conservation $[H,L^2] = 0$ implies that $\mathcal{H}_l$ is stable under the action of H. We here assume that H on $\mathcal{H}_l$ has the same properties as on $\mathcal{H}$ so that the decompositioin (4-1) is also valid for $\mathcal{H}_l$.

The above assumption also implies that $[S,L^2] = 0$. Thus the S operator leaves the subspaces $\mathcal{H}_l$ invariant. We call S_l the restriction of S on $\mathcal{H}_l$. If we assume asymptotic completeness, S_l is unitary on $\mathcal{H}_l$. Within $\mathcal{H}_l$, (2-1) reads:

$$f_l^{out} = S_l \, g_l^{in} \quad ; \quad f_l^{out}, g_l^{in} \in \mathcal{H}_l \tag{4-4}$$

This situation can be achieved if the potential V is spherically symmetric in the Schrodinger representation.

5. We call the realization of the space of states $\mathcal{H}$ by functions of the energy (the values of these functions being the components of the vector states with respect to the generalized basis system which diagonalizes H_o) the energy representation. In this realization $f^{out} = S\, g^{in}$ reads

$$f^{out}(E) = S(E)g^{in}(E) \tag{4-5}$$

and the S matrix, S(E), is in general an infinite matrix and its elements are complex functions of E. Equation (4-4) in the energy representation reads:

$$f_l^{out}(E) = S_l(E) \, g_l^{in}(E) \tag{4-5-A}$$

For $l = 0$, $S_o(E)$ is a complex function of the positive variable E. $S_o(E)$ can be extended to be a meromorphic (analytic except for isolated singularities) complex function on a two sheeted Riemann surface (corresponding to the transformation $E = P^2/2m$ where E and P are complex numbers[2]). We denote this extension into the complex plane by $S_o(\omega)$. Since we shall treat only the case of $l = 0$ in this section, we hereafter omit the subscript o unless it would cause confusion, and write $S(\omega)$ and $S(E)$ for $S_o(\omega)$ and $S_o(E)$, respectively.

$S(\omega)$ has in general an infinite number of simple poles on the second sheet. No other kind of singularities will exist there (although we may allow simple poles on the negative real axis of the first sheet, which correspond to bound states). The poles on the second sheet occur either in pairs at locations that are complex conjugate of each other (resonances) or are located on the negative real axis (virtual states). For the sake of simplicity, we first restrict ourselves to the simplest case of one pair of resonance poles. These poles are assumed to be located at

$$\omega = E_R \pm i\Gamma/2 \tag{4-6}$$

where Γ and E_R are positive numbers representing the width and resonance energy respectively. The extention to the case of a finite number of pairs of poles (with imaginary parts different from zero) is straightforward.

At this point, we recall the structure of the Riemann surface associated with the transformation z^2. It consists of two complex planes having the origin in common. These planes are cut along the positive real axis; the lower edge of this cut in the second plane is glued together with the upper edge of the first plane and the lower edge of the first plane is glued to the upper edge of the second plane. The resulting surface (Riemann Surface) consists of two sheets which intersect each other along the positive real axis, crossing the real axis one gets from the first sheet into the second sheet and vice versa.

A neighborhood of a point on the upper rim contains points of the upper half plane of the first sheet and points in the lower half plane of the second sheet. A neighborhood of a point in the lower rim contains points in the lower half plane of the first sheet and points in the upper half plane of the second sheet. A neighborhood of zero contains points in both sheets and has a "snaillike" form as shown in Figure 5 . For more details see Reference 3 at the end of this chapter.

6. We have to make some assumptions concerning the behavior of $S(\omega)$ at infinite on the second sheet. This is needed in our derivation in order that certain integrals vanish. Our hypothesis is that $S_{II}(\omega)$ (values of $S(\omega)$ on the second sheet) is polynomially bounded at infinite, i.e.,

$$|S_{II}(\omega)| \leq |P(\omega)| \quad , \tag{4-7}$$

where $P(\omega)$ is a polynomial, for any $\omega \in C$ whose modulus is larger than some $R > 0$.

Now, let φ^{out} be any outgoing free state and ψ^{in} any incoming free state (cf. Chapter II, Section I) and S the scattering operator on H_0 (we restrict ourselves to $l = 0$ here). We consider the matrix element:

$$(\varphi^{out}, S\psi^{in}) , \quad \varphi^{out}, \psi^{in} \in H_0 \ . \tag{4-8}$$

One can easily prove that the scalar product in (4-8) does not depend on time t. If $\varphi^{out}(t)$ and $\psi^{in}(t)$ are defined by $\varphi^{out}(t) = e^{-itH_0} \varphi^{out}$ and $\psi^{in}(t) = e^{-itH_0} \psi^{in}$:

$$(\varphi^{out}(t), S\psi^{in}(t)) = (\varphi^{out}, S\psi^{in}) \ . \tag{4-9}$$

On the other hand, if we use (4-2) we obtain:

$$(\varphi^{out}, S\psi^{in}) = (\varphi^{out}, (\Omega^-)^* \Omega^+ \psi^{in}) = (\Omega^- \varphi^{out}, \Omega^+ \psi^{in}) = (\varphi^-, \psi^+) \tag{4-10}$$

where $\varphi^- = \Omega^-\varphi^{out}$, $\psi^+ = \Omega^+\psi^{in}$ are the interacting state vectors corresponding asymptotically ($t \to \pm \infty$) to the non-interacting out- and in-states respectively. We have to choose $\varphi^- \in \Phi^-$ and $\psi^+ \in \Phi^+$. Then $\varphi^{out} \in \Phi_-$ and $\psi^{in} \in \Phi_+$.

Now consider

$$(\varphi^{out}, S\psi^{in}) = (U\varphi^{out}, USU^{-1}U\psi^{in}) \qquad (4\text{-}11)$$

where U is the unitary operator in the diagram of Chapter III.

$U\varphi^{out} \in D_+$. It is therefore a function of E. As a consequence of the Gelfand Maurin theorem, $U\varphi^{out} = \langle E_- | \varphi^{out} \rangle$ (see considerations on Chapter III and formula (3-54)). Also, $U\psi^{in} \in D_-$ and $U\psi^{in} = \langle E_+ | \psi^{in} \rangle$.

For $l = 0$; $USU^{-1} = S(E)$ is a complex function of E for $E > 0$. E represents here the physical values of the energy. Therefore, the values of S(E) can be identified with the boundary values of the S-matrix $S(E + i0)$ along the upper rim of the cut.[4)] These ideas lead to the following formula[5)]:

$$(\varphi^{out}, S\psi^{in}) = \int_0^\infty \langle \varphi^{out} | E_- \rangle \langle E_+ | \psi^{in} \rangle \, S(E + i0) \, dE \qquad (4\text{-}12)$$

where the functions in the integrand are defined on the upper rim of the cut.

Since $\varphi^{out} \in \Phi_-$ and $\psi^{in} \in \Phi_+$, it follows that $\langle \varphi^{out} | E_- \rangle$ and $\langle E_+ | \psi^{in} \rangle = \langle \psi^{in} | E_+ \rangle^*$ belong to D_- and are, therefore, boundary values on R^+ of functions in H_-.

We have already mentioned that a neighbourhood of a point on the upper rim contains points of the upper half plane of the first sheet (also called the physical sheet) and points in the lower half plane of the second sheet. Therefore, the upper rim connects these two half planes. For the same reason, the functions in the integrand of (4-12) have their analytic continuations on the lower half plane of the second sheet.

To continue the calculation of (4-12) we use

$$\langle \varphi^{out} | E_- \rangle = \langle \Omega^-\varphi^{out} | (\Omega^-)^{\times} | E_- \rangle = \langle \varphi^- | E^- \rangle \qquad (4\text{-}13\text{-}A)$$

and

$$\langle E_+ | \psi^{in} \rangle = \langle \psi^{in} | E_+ \rangle^* = \langle \Omega^+\psi^{in} | (\Omega^+)^{\times} | E_+ \rangle^* = \langle \psi^+ | E^+ \rangle^* = \langle E^+ | \psi^+ \rangle \; . \qquad (4\text{-}13\text{-}B)$$

Inserting this and (4-9) into (4-12), yields

$$(\varphi^-,\psi^+) = \int^\infty_0 S(E+i0)\langle\varphi^-|E^-\rangle\langle^+E|\psi^+\rangle\, dE \ . \qquad (4\text{-}14)$$

We have assumed that $S(\omega)$ has a pair of simple poles on the second sheet. Then, the analytic function, whose boundary values on the upper rim of the cut is the integrand in (4-14), has a simple pole on the lower half plane of the second sheet. For $\mathrm{Im}\omega > 0$, the values of this function are

$$S_{II}(\omega)\,\langle\varphi^-|\omega^-\rangle\langle^+\omega^*|\psi^+\rangle \qquad (4\text{-}15)$$

where $|\omega^\pm\rangle$ were defined in (3-77).

The pole of $S_{II}(\omega)$ and therewith of (4-15) is at

$$\omega = Z_R \equiv E_R - i\Gamma/2 \ . \qquad (4\text{-}16)$$

By an elementary application of the residue theorem, we have

$$\int_C S_{II}(\omega)\,\langle^-\varphi|\omega^-\rangle\langle^+\omega^*|\psi^+\rangle\, d\omega = 2\pi i\, \mathrm{Res}[(4\text{-}15), Z_R] \qquad (4\text{-}17)$$

where C is the contour depicted in Figure 6. Res $[(4\text{-}15), Z_R]$ is the residue of (4-15) at the point Z_R, where the pole is located. C is the union of a semicircle and a segment. When the radius of the semicircle goes to infinite, the integral over it vanishes, due to the condition (4-6).[6] The segment becomes a straight line of infinite length, which can be pulled toward the real axis (in the second sheet), whose positive half includes the upper rim of the cut. Then, (4-17) becomes

$$\begin{aligned} -2\pi i\, \mathrm{Res}[(4\text{-}15), Z_R] &= \int^0_{-\infty} S_{II}(E)\,\langle^-\varphi|E^-\rangle\langle^+E|\psi^+\rangle\, dE \\ &+ \int^\infty_0 S(E+i0)\,\langle^-\varphi|E^-\rangle\langle^+E|\psi^+\rangle\, dE \end{aligned} \qquad (4\text{-}18)$$

where $E \in R^-$ (second sheet) in the first integral and $E \in$ {continuous spectrum of H} in the second integral of (4-18).

$|E^\pm\rangle$ for $E \in R^-$ are defined exactly as $|\omega^\pm\rangle$ (see previous chapter). However, their continuity cannot be proven exactly as we did in the case of $|\omega^\pm\rangle$ for Im $\omega \neq 0$.

In fact, if $g(E) = \theta^{-1}U\varphi_\pm$ and we choose $E \in R^-$, (3-67) gives

$$\langle\varphi^\pm|E^\pm\rangle = \langle\varphi_\pm|E_\pm\rangle = g^*(E) = \int^\infty_{-\infty} g(E')\delta^*(E'-E)\, dE' = \langle g(E')|\delta^*(E'-E)\rangle \qquad (4\text{-}19)$$

$\delta^*(E'-E)$ is a functional on S which carries a function $f(E) \in S$ to the complex conjugate of its value at E. It is well known that $\delta^*(E'-E)$ is a continuous antilinear functional on S. On the other hand, it is not difficult to prove that

$$\Delta\pm \subset S \subset L^2(R) \subset S^x \subset (\Delta_\pm)^x \ . \qquad (4\text{-}20)$$

Hence $\delta^*(E' - E) \in (\Delta_\pm)^x$ and from (4-19) we have

$$|E_\pm\rangle = (U^{-1})^x\,\theta^x\,\delta^*(E' - E)\ . \qquad (4\text{-}21)$$

From (4-21), it follows that $|E_\pm\rangle \in (\Phi_\pm)^x$ if $E \in R^-$. Also, $|E^\pm\rangle = (\Omega^\pm)^x|E_\pm\rangle \in (\Phi^\pm)^x$.

$$(\varphi^-,\psi^+) = \int^{-\infty}{}_0\, S_{II}(E)\,\langle^-\varphi|E^-\rangle\langle^+E|\psi^+\rangle\, dE - 2\pi i\, \mathrm{Res}\{(4\text{-}15), Z_R]\ . \qquad (4\text{-}22)$$

We call the first term on the righthand side of (4-21) the background term.

In the neighbourhood of the pole Z_R the S-matrix can be expanded into a Laurent series:

$$S_{II}(\omega) = s_{-1}(\omega - Z^R)^{-1} + s_o + s_1(\omega - Z_R) + \dots \qquad (4\text{-}23)$$

If we use the definition of the residue[6)] and the fact that $\langle^-\varphi|\omega^-\rangle\langle^+\omega^*|\psi^+\rangle$ is analytic for $\mathrm{Im}\,\omega > 0$, then we have

$$\mathrm{Res}[(4\text{-}15), Z_R] = s_{-1}\langle^-\varphi|Z_R\rangle\langle^+Z_R{}^*|\psi^+\rangle\ . \qquad (4\text{-}24)$$

Using (4-24), one obtains for (4-22):

$$(\varphi^-,\psi^+) = \text{background} - 2\pi i\, s_{-1}\langle^-\varphi|Z_R{}^-\rangle\langle^+Z_R{}^*|\psi^+\rangle\ . \qquad (4\text{-}25)$$

We now start again from formula (4-14) and take its complex conjugate:

$$(\psi^+,\varphi^-) = \int^\infty{}_0\, S^*(E + i0)\langle^+\psi|E^+\rangle\langle^-E|\varphi^-\rangle\, dE\ . \qquad (4\text{-}26)$$

Then, we use (2-8) to obtain

$$(\psi^+,\varphi^-) = \int^\infty{}_0\, S(E - i0)\langle^+\psi|E^+\rangle\langle^-E|\varphi^-\rangle\, dE\ . \qquad (4\text{-}27)$$

S(E - i0) is the boundary value on the lower rim of the cut of an analytic function on the upper half plane of the second sheet having a simple pole at[2-3)]

$$Z_R^* = E_R + i\Gamma/2 \ . \tag{4-28}$$

For $E \in R^+$, $\langle^-\varphi|E^-\rangle$ and $\langle^+E|\psi^+\rangle$ are functions in D_- and their respective complex conjugates $\langle^-E|\varphi^-\rangle$ and $\langle^+E|\psi^+\rangle$ are in D_+. To obtain the value of (4-27), we define the function $\langle^+\psi|E^+\rangle\langle^-E|\varphi^-\rangle$ on the lower rim. Then, the integrand in (4-27) makes sense on this lower rim, it is the boundary value of a function on the upper half plane on the second sheet analytic except for a simple pole at Z_R^*. This function is ($\mathrm{Im}\,\omega > 0$):

$$S_{II}(\omega)\,\langle^+\psi|\omega^{*+}\rangle\langle^-\omega|\varphi^-\rangle \ . \tag{4-29}$$

By contour integration over C' (see Figure 7) on the upper half plane of the second sheet, we have:

$$\int_{C'} S_{II}(\omega)\,\langle^+\psi|\omega^{*+}\rangle\langle^-\omega|\varphi^-\rangle\, d\omega = 2\pi i\ \mathrm{Res}[(4\text{-}29), Z_R^*] \ . \tag{4-30}$$

C' is the union of a semicircle plus a straight segment parallel to the real axis. We can make the radius of the semicircle go to infinite and pull the segment toward the real axis, which is here the negative real axis of the second sheet plus the lower rim of the cut. At the limit, the integral over the semicircle will vanish because of (4-7).[6)] The integral over the real axis becomes:

$$\int^{0}_{-\infty} S_{II}(E)\langle^+\psi|E^+\rangle\langle^-E|\varphi^-\rangle\, dE \ + \ \int^{\infty}_{0} S(E - i0)\,\langle^+\psi|E^+\rangle\langle^-E|\varphi^-\rangle\, dE$$
$$= 2\pi i\ \mathrm{Res}[(4\text{-}29), Z_R^*] \tag{4-31}$$

where $E \in R^-$ in the first integral. The negative of the first integral we call background.* If σ_{-1} denotes the residue of $S_{II}(\omega)$ at Z_R^*, then $\sigma_{-1}\langle^+\psi|Z_R^{*+}\rangle\langle^-Z_R|\varphi^-\rangle$ is the residue of (4-29) at Z_R^*. With (4-27) we thus have:

$$(\psi^+,\varphi^-) = \mathrm{background}^* + 2\pi i\sigma_{-1}\,\langle^+\psi|Z_R^{*+}\rangle\langle^-Z_R|\varphi^-\rangle \ . \tag{4-32}$$

This is the analogue of (4-25).

If in (4-25) we omit the arbitrary $\varphi^- \in \Phi^-$, we can write formally:

$$\psi^+ = \int^{-\infty}_{0} S_{II}(E)|E^-\rangle\langle^+E|\psi^+\rangle\, dE - 2\pi i\ s_{-1}|Z_R^-\rangle\langle^+Z_R^*|\psi^+\rangle \ . \tag{4-25-A}$$

If in (4-32) we omit the arbitrary $\psi^+ \in \Phi^+$, we have:

$$\varphi^- = \int^{-\infty}_{0} S_{II}(E)\ |E^+\rangle\langle^-E|\varphi^-\rangle\, dE + 2\pi i\ \sigma_{-1}\ |Z_R^{*+}\rangle\langle^-Z_R|\varphi^-\rangle \ . \tag{4-32-A}$$

The integral term in (4-25-A) represents a functional on Φ^-. If we apply this functional to any $\varphi^- \in \Phi^-$ the result we obtain is the background term in (4-25). It can be written as a linear combination of ψ^+ and $|Z_R^-\rangle$. $\psi^+ \in \Phi^+ \subset \mathcal{H}_{ac} \subset (\Phi^-)^\times$ and $|Z_R^-\rangle$ is obviously in $(\Phi^-)^\times$. Therefore this integral term represents a continuous antilinear functional on Φ^-. Similar arguments will lead to the conclusion that the integral term in (4-32-A) is a functional in $(\Phi^+)^\times$.

Finally, as a consequence of (3-78), we have

$$H_1^\times |Z_R^-\rangle = Z_R |Z_R^-\rangle \tag{4-33-A}$$

$$H_1^\times |Z_R^{*+}\rangle = Z_R^* |Z_R^{*+}\rangle \ . \tag{4-33-B}$$

The vectors $|Z_R^-\rangle$ and $|Z_R^{*+}\rangle$ associated with the resonance pole at $Z_R = E - i\,\Gamma/2$ and $Z_R^* = E + i\Gamma/2$ are generalized eigenvectors of the self-adjoint Hamiltonian with complex eigenvalue Z_R and Z_R^*, respectively.

II. DYNAMICAL SEMIGROUPS

Our next goal is to study the time evolution of the vectors $|Z_R^-\rangle$ and $|Z_R^{*+}\rangle$. This study must obviously start with a definition of time evolution for the members of the (anti)duals. Let us recall briefly the corresponding idea for vectors in Hilbert space. If the vector state $\varphi \in H$ undergoes the action of a force given by a time independent potential V, its time evolution is provided by the formula:

$$\varphi(t) = e^{-itH}\varphi \quad \text{with} \quad H = H_o + V . \tag{4-34}$$

As we have already mentioned (see Section II in Chapter II), if φ is also $\in H_{ac}$ then:

$$e^{-itH}\varphi = e^{-itH_1}\, \varphi , \tag{4-35}$$

the time evolution of φ is regulated by the continuous part H_1 of the total hamiltonian.

On the other hand, if U is a unitary operator on H and $\Phi \subset H \subset \Phi^x$ is a R.H.S., we can extend U to Φ^x provided that

1. The Hilbert space adjoint of U, U^+, leave Φ invariant: $U^+\Phi \subset \Phi$.
2. U^+ is continuous on Φ with respect to the nuclear topology on Φ.

We call U^x the extension of U to Φ^x. It is defined by:

$$\langle U^+\varphi|F\rangle = \langle \varphi U^x F\rangle = \quad \forall\, \varphi \in \Phi ; \quad F \in \Phi^x \tag{4-36}$$

Since Φ is stable under the action of U^+, $U^+\varphi \in \Phi$ $\forall \varphi \in \Phi$ and theleft hand side of (4-36) makes sense. The reader can easily verify that

1. U^xF is an antilinear functional on Φ for any $F \in \Phi^x$, i.e., $U^x\Phi^x \subset \Phi^x$.
2. U^xF is an extension of U defined on H.

Let $F \in H$ and let φ be any element of Φ:

$$\langle \varphi|U^xF\rangle = \langle U^+\varphi|F\rangle = (U^+\varphi,F) = (\varphi,UF) = \langle \varphi|UF\rangle \tag{4-37}$$

and hence $U^xF = UF$.

3. U^xF belongs to Φ^x or equivalently U^xF is continuous.

Since we are only considering spaces Φ that fulfill the first axiom of countability, to prove the above assertion we need to show that from

$\varphi_n \to \varphi$ with τ_Φ

follows

$\langle\varphi_n|U^xF\rangle \to \langle\varphi|U^xF\rangle$ (4-38-A)

or equivalently

$\langle U^+\varphi_n|F\rangle \to \langle U^+\varphi|F\rangle$. (4-38-B)

We have assumed that U^+ is continuous on Φ, which implies that if

$\varphi_n \to \varphi$ with τ_Φ

then

$U^+\varphi_n \to U^+\varphi$ with τ_Φ . (4-39)

Hence, since F is continuous on Φ, (4-38-B) holds and the continuity of U^xF on Φ is proven.

Finally, we want to assert that U^x is a continuous operator on Φ^x. We are not proving this statement here. It in particular implies that if $F_n \to F$, then $U^xF_n \to U^xF$.

The next result is of great importance.

<u>Propositiion I</u>.

If $t > 0$ $e^{-itH_1}\ \Phi^- \subset \Phi^-$ (4-40-A)

If $t > 0$ $e^{-itH_1}\ \Phi^+ \subset \Phi^+$ (4-40-B)

However

If $t > 0$ $e^{-itH_1}\ \Phi^- \not\subset \Phi^-$ (4-41-A)

If $t > 0$ $e^{-itH_1}\ \Phi^+ \not\subset \Phi^+$ (4-41-B)

<u>Proof</u>. If we prove that

$e^{itE}D_+ \subset D_+$ for $t > 0$ (4-42-A)

we would have proven (4-40-A). Also if we prove

$e^{itE}D_- \subset D_-$ for $t > 0$ (4-42-B)

we would have proven (4-40-B). The reason for this is as follows: Assume (4-42-A) holds. Then for any $t > 0$,

$\Omega^-U^{-1}\,e^{itE}D_+ \subset \Omega^-U^{-1}D_+ = \Omega^-\Phi_- = \Phi^-$ (4-43)

where U is the operator defined in (3-11) . If we insert (4-43), we obtain

$\Omega^-U^{-1}\,e^{itE}\,UU^{-1}D_+ = \Omega^-\,e^{itH_0}\,U^{-1}D_+ = \Omega^-\,e^{itH_0}\,\Phi_-$. (4-44)

This procedure is correct because, since E is e.s.a., $U^{-1}\,EU = H_0$ implies $U^{-1}\,e^{itE}\,U = e^{itH}$ and vice versa for any real t. Now, if we insert $I = (\Omega^-)^{-1}\,\Omega^-$ in the right hand side of (4-44):

$$\Omega^-\,e^{itH_0}\,\Phi_- = \Omega^-\,e^{itH_0}\,(\Omega^-)^{-1}\,\Omega^-\,\Phi_-$$
$$= e^{itH_1}\,\Omega^-\Phi_- = e^{itH_1}\Phi^- \ . \quad (4\text{-}45)$$

From all of the above, we conclude (4-40-A). The relation between (4-42-B) and (4-40-B) can be demonstrated in the same way. It remains to prove (4-42).

We are still assuming that $t > 0$. Pick $\varphi_+(E) \in D_+$. Then, $\theta^{-1}\varphi_+(E) = g_+(E) \in \Delta_+$ according to (3.21). Consider $e^{itE}g_+(E)$. Since e^{itE} for any fixed $t \in R$ is a bounded function having continuous bounded derivatives to all orders, one can easily prove that $e^{itE}g_+(E) \in S$ as a function of $E \in R$ for any t. If we are able to prove that for $t > 0$ $e^{itE}g_+(E) \in H_+$, then $e^{itE}g_+(E) \in \Delta_+$ and hence

$$\theta e^{itE}g_+(E) = e^{itE}\theta g_+(E) = e^{itE}\varphi_+(E)$$
$$= e^{itE}\varphi_+(E) \in D_+ \quad \text{for} \quad t>0 . \tag{4-46}$$

Consequently, (4-42-A) would follow.

To justify that $f_+(E) = e^{itE}g_+(E) \in H_+$ for $t > 0$, we have to show that if $\omega = E+ i\delta$, $\delta > 0$, then

$$\int^{\infty}_{-\infty} |f_+(E + i\delta)|^2 \, dE < k < \infty \qquad k \text{ fixed} \qquad \forall\, \delta > 0 \tag{4-47}$$

because of the definition of Hardy class functions on the upper half plane. Simple manipulations give

$$f_+(E+ i\delta) = e^{it(E+i\delta)}g_+(E+ i\delta) = e^{itE}\, e^{-t\delta}g_+(E+i\delta) . \tag{4-48}$$

Taking moduli

$$|f_+(E+ i\delta)| = |e^{-t\delta}\, g_+(E+ i\delta)| = e^{-t\delta}|g_+(E+ i\delta)| \le |g_+(E+ i\delta)| . \tag{4-49}$$

The inequality in (4-49) is correct because $\delta > 0$ and $t > 0$. Therefore, since $g_+(E) \in H_+$ and (4-49)

$$\int^{\infty}_{-\infty} |f_+(E+ i\delta)|^2 dE \le \int^{\infty}_{-\infty} |g_+(E+ i\delta)|^2 \, dE < k < \infty ; \quad \forall\, \delta > 0 \tag{4-50}$$

whence (4-47) follows. Similar manipulations prove (4-42-B) and hence (4-40-B). We do not supply here the proofs of (4-41). The interested reader can find them in Reference 5.

Nevertheless, we make here a couple of remarks regarding these formulas.

(4-41-A) says that (4-40-A) is no longer valid for $t < 0$ and (4-41-B) says that (4-40-B) is no longer valid for $t > 0$. As was the case for (4-40), we shall have shown (4-41-A), if we can demonstrate that

$$e^{it\hat{E}}D_+ \not\subset D_+ \qquad t<0 \tag{4-51-A}$$

and (4-41-B), if

$$e^{it\hat{E}}D_- \not\subset D_- \qquad t>0 . \tag{4-51-B}$$

For (4-51-A) we have to show that for any $t < 0$, there exists <u>some</u> $\varphi_+(E) \in D_+$ such that $e^{itE}\varphi_+(E)$ is not in D_+ or equivalently that $\theta^{-1} e^{itE}\varphi_+(E) = e^{itE}\, \theta^{-1}\varphi_+(E)$ is not in Δ_+. For

(4-50-B) we have to find for any $t > 0$ some $\psi_-(E) \in D_-$ such that $e^{itE}\psi_-(E)$ is not in D_-. This is done in Reference 5.

Proposition I is one of the central results of the chapter. We give two more results here.

Proposition II. Consider the Schwartz space S. Let $\hat{\varepsilon}_t$ be an operator on S such that if $f(E) \in S$

$$\hat{\varepsilon}_t f(E) = e^{itE} f(E) \ . \tag{4-52}$$

Then, $\hat{\varepsilon}_t$ is continuous on S for all $t \in R$.

Proof: See Appendix VI.

Proposition III: e^{itH_1} is a continuous operator on Φ^- if $t > 0$ and on Φ^+ if $t < 0$.

Proof. We first need to show that

$$e^{it\hat{E}} \text{ is continuous on } D_+ \text{ for } t > 0 \tag{4-53-A}$$

$$e^{it\hat{E}} \text{ is continuous on } D_- \text{ for } t < 0 \ . \tag{4-53-B}$$

Assume $\varphi_n(E) \to \varphi(E)$ in D_+. Then, if $g_n(E) = \theta^{-1}\varphi_n(E)$ and $g(E) = \theta^{-1}\varphi(E)$, $g_n(E) \to g(E)$ in Δ_+ and hence in S. After Proposition II, $e^{itE} g_n(E) \to e^{itE} g(E)$ in S. If $t > 0$ $e^{itE} g_n(E)$ and $e^{itE} g(E)$ are in Δ_+. Therefore,

$$\theta e^{itE} g_n(E) \to \theta e^{itE} g(E) \text{ in } D_+ \tag{4-54-A}$$

$$\| \qquad\qquad \|$$

$$e^{itE}\theta g_n(E) \to e^{itE}\theta g(E) \tag{4-54-B}$$

$$\| \qquad\qquad \|$$

$$e^{itE}\varphi_n(E) \to e^{itE}\varphi(E) \tag{4-54-C}$$

$$\| \qquad\qquad \|$$

$$e^{it\hat{E}}\varphi_n(E) \to e^{it\hat{E}}\varphi(E) \tag{4-54-D}$$

which proves(4-53-A).We obtain(4-53-B) by using the same ideas and replacing D_+ by D_-.

If we now apply U^{-1} to (4-54-D)

$$U^{-1}e^{it\hat{E}}\varphi_n(E) \to U^{-1}e^{it\hat{E}}\varphi(E) \text{ in } \Phi_- \tag{4-55-A}$$

$$\| \qquad\qquad \|$$

$$U^{-1}e^{it\hat{E}}UU^{-1}\varphi_n(E) \to U^{-1}e^{it\hat{E}}UU^{-1}\varphi(E) \tag{4-55-B}$$

$$\| \qquad\qquad \|$$

$$e^{itH_o}U^{-1}\varphi_n(E) \to e^{itH_o}U^{-1}\varphi(E) \ . \tag{4-55-C}$$

If we multiply to the left by Ω^-, we have

$$\Omega^- e^{itH_o} U^{-1}\varphi_n(E) \to \Omega^- e^{itH_o} U^{-1}\varphi(E) \in \Phi^- \tag{4-56-A}$$

$$\| \qquad\qquad\qquad \|$$

$$\Omega^- e^{itH_o} (\Omega^-)^{-1}\Omega^- U^{-1}\varphi_n(E) \to \Omega^- e^{itH_o}(\Omega^-)^{-1}\Omega^- U^{-1}\varphi(E) \tag{4-56-B}$$

$$\| \qquad\qquad\qquad \|$$

$$e^{itH_1} \Omega^- U^{-1}\varphi_n(E) \to e^{itH_1} \Omega^- U^{-1}\varphi(E) \ . \tag{4-56-C}$$

$\varphi_n(E)$ is an arbitrary sequence in D_+ and $\Omega^- U^{-1}$ a bicontinuous bijection from D_+ onto Φ^-. Thus, $\varphi^-_n = \Omega^- U^{-1}\varphi_n(E)$ is an arbitrary convergent sequence in Φ^-. Its limit is $\varphi^- = \Omega^- U^{-1}\varphi(E)$. Therefore, (4-56-C) says that if

$$\varphi^-_n \to \varphi^- \quad \text{in} \quad \Phi^-$$

then

$$e^{itH_1} \varphi^-_n \to e^{itH_1} \varphi^- \quad \text{in} \quad \Phi \tag{4-57}$$

if $t > 0$, and consequently e^{itH_1} is continuous on Φ^- if $t > 0$. The proof for Φ^+ and $t < 0$ is identical.

As a direct consequence of Proposition III we have:

1. For $t > 0$, $(e^{-itH_1})^x$ can be defined on $(\Phi^-)^x$ because the Hilbert space adjoint of e^{-itH_1}, e^{itH_1}, leaves Φ^- invariant. Moreover, $(e^{-itH_1})^x$ is continuous on $(\Phi^-)^x$.
2. For $t < 0$, $(e^{-itH_1})^x$ is a continuous operator on $(\Phi^+)^x$.

Proof: Recall (4-36) and (4-38).

We are now in the position to define the time evolution for the functionals $F^\pm(\omega) = |\omega^\pm\rangle \in (\Phi^\pm)^x$ [and, in particular, for $|Z_R^-\rangle$ and $|Z^{*+}_R\rangle$]. $F^+ \in (\Phi^+)^x$ will evolve in time under the action of $(e^{-itH_1})^x$ when $t < 0$ and $F^- \in (\Phi^-)^x$ when $t > 0$. As a consequence of (4-41), if $t > 0$ ($t < 0$) no evolution of this kind can be defined on $F^+ \in (\Phi^+)^x$ ($F^- \in (\Phi^-)^x$).

For the time $t > 0$ we obtain from $F^-(0) = F^-$, the functional

$$F^-(t) = (e^{-itH_1})^x F^- \quad (t > 0) \tag{4-58-A}$$

and for the time $t < 0$ the functional $F^+(t)$ is given in terms of $F^+(0) = F^+$ by

$$F^+(t) = (e^{-itH_1})^x F^+ \quad (t < 0) \ . \tag{4-58-B}$$

In particular, if $F^-(0) = |Z_R^-\rangle$, after some time $t > 0$, this state vector becomes $(e^{-itH_1})^x|Z_R^-\rangle$, as follows from the formula

$$\langle e^{itH_1}\varphi^- | Z_R^- \rangle = \langle \varphi^- | (e^{-itH_1})^x | Z_R^- \rangle \qquad \forall \varphi^- \in \Phi^- ; \quad t > 0 . \qquad (4\text{-}59\text{-}A)$$

If $F^+(0) = |Z^{*+}_R\rangle$, at some instant t in the past ($t < 0$), this state was given by $(e^{-itH_1})^x |Z^{*+}_R\rangle$ as follows from

$$\langle e^{-itH_1}\varphi^+ | Z^{*+}_R \rangle = \langle \varphi^+ | (e^{-itH_1})^x | Z^{*+}_R \rangle \qquad \forall \psi^+ \in \Phi^+ ; \quad t < 0 . \qquad (4\text{-}59\text{-}B)$$

(4-59) is a particular form of (4-36).

The next result is the most relevant one and gives us a motivation for the above manipulations

Proposition IV.

$$(e^{-itH_1})^x |Z_R^-\rangle = \exp\{-itZ_R\} |Z_R^-\rangle ; \quad t > 0 \qquad (4\text{-}60\text{-}A)$$

$$(e^{-itH_1})^x |Z_R^{*+}\rangle = \exp\{-itZ_R^*\} |Z_R^{*+}\rangle ; \quad t < 0 . \qquad (4\text{-}60\text{-}B)$$

Proof: Take $t > 0$ and apply the left hand side of (4-60-a) to an arbitrary $\varphi^- \in \Phi^-$. Consider $\langle \varphi^- | Z_R^- \rangle = [g_+(Z_R)]^*$, where $\theta^{-1}U(\Omega^-)^{-1}\varphi^- = g_+(E) \in \Delta_+$. $[g_+(Z_R)]^*$ is the value of $[g_+(\omega)]^* \in H_-$ at the point Z_R. This value is equal to $[g_+(Z_R^*)]^*$. If $t > 0$ then $\theta^{-1}U(\Omega^-)^{-1} e^{itH_1}\varphi^- = e^{itE}g_+(E) \in \Delta_+$. Thus

$$\begin{aligned} \langle \varphi^- | (e^{-itH_1})^x | Z_R^- \rangle &= \langle e^{itH_1}\varphi^- | Z_R^- \rangle \\ &= [\exp\{itZ_R^*\}\, g_+(Z_R^*)]^* = \exp\{-ittZ_R\}[g_+(Z_R^*)] \\ &= \exp\{-itZ_R\}\, [g_+(Z_R)]^* \qquad \text{(see Ref. 9 of Chapter III)} \\ &= \exp\{-itZ_R\}\, \langle \varphi^- | Z_R^- \rangle \\ &= \langle \varphi^- | \exp\{-itZ_R\} | Z_R^- \rangle . \end{aligned} \qquad (4\text{-}61)$$

Take the first and last term in (4-6l) and omit the arbitrary $\varphi^- \in \Phi^-$ to obtain (4-60-A). The proof of (4-60-B) is identical.

Corollary. Since $Z_R = E_R - i\Gamma/2$, we obtrain as a consequence of (4-60-A):

$$(e^{-itH_1})^x |Z_R^-\rangle = \exp\{-itE_R\}\, e^{-\Gamma t/2} |Z_R^-\rangle ; \quad t > 0 . \qquad (4\text{-}62\text{-}A)$$

Thus $|Z_R^-\rangle$ decays exponentially starting at the time $t = 0$.

Since $Z_R^* = E_R + i\Gamma/2$ we obtain as a consequence of (4-60-B):

$$(e^{-itH_1})^x |Z_R^{*+}\rangle = \exp\{-itE_R\}\, e^{\Gamma t/2} |Z_R^{*+}\rangle \quad t < 0 . \qquad (4\text{-}63\text{-}A)$$

Thus $|Z_R{}^{*+}\rangle$ has grown exponentially and has become $|Z_R{}^{*+}\rangle$ at the time $t = 0$.

We want to emphasize that (4-60-A) and (4-62-A) are not valid for $t < 0$ and that (4-60-B) and (4-62-B) are not valid for $t > 0$.

<u>A remark concerning terminology:</u> Henceforth, we call $|Z_R{}^{-}\rangle$ the decaying Gamow vector and $|Z_R{}^{*+}\rangle$ the growing Gamow vector.

Proposition IV uncovers the physical meaning of equatioons (4-24-A) and (4-31-A). In (4-24-A), ψ^+ represents a decaying state. ψ^+ is a linear combination of a purely exponential decaying state, the Gamow vector $|Z_R{}^{-}\rangle$ plus the background. It starts to decay at $t = 0$. In (4-31-A), φ^- represents a growing state. It is a linear combination of a purely exponential growing state, the Gamow vector $|Z_R{}^{*+}\rangle$, plus again a term called background. φ^- stops to grow at $t = 0$.

What is the physical interpretation of the term? If we perform a decay experiment, we shall observe a purely exponential decay law. However, if we study the decay from a theoretical point of view in the Hilbert space formulation of quantum mechanics deviations from this exponential law are derived.[8)] The deviations are stronger for positive small times and for very large ones so that on these ranges the exponential behavior is overwhelmed by these deviations. The usual physical interpretation of these mathematical results is caused by the interactions of the decaying state with the environment and the measuring apparatus. In our formulation of quantum mechanics, the exponentially decaying part, $|Z_R{}^{-}\rangle$, is separated from the background term and the background term accounts for the deviations from the exponential law. At $t = \infty$, the decaying state has completely decayed. But also the background term goes to zero for $t \to \infty$. This can be formally expressed as:

$$\lim \ (e^{-itH_1})^x \int_0^{-\infty} |E^-\rangle\langle^+E|\psi^+\rangle\, S_{II}(E)\, dE = 0 \ ; \ \text{as } t \to 0 \qquad (4\text{-}64)$$

To prove (4-64), take $t > 0$, $\varphi^- \in \Phi^-$ and E real. Then

$$\begin{aligned}
&\langle\varphi^-|(e^{-itH_1})^x|E^-\rangle = \langle e^{itH_1}\varphi^-|E^-\rangle \\
&= [e^{itE} g_+(E)]^* = e^{-itE}[g_+(E)]^* \\
&= e^{-itE}\langle\varphi^-|E^-\rangle = \langle\varphi^-|e^{-itE}|E^-\rangle \ .
\end{aligned} \qquad (4\text{-}65)$$

This implies that

$$(e^{-itH_1})^x|E^-\rangle = e^{-itE}|E^-\rangle \qquad (4\text{-}66)$$

for any real E, positive or negative.

$$\begin{aligned}
&\langle\varphi^-|(e^{-itH_1})^x|\text{background}\rangle \\
&= \langle e^{itH_1}\varphi^-|\text{background}\rangle \\
&= \int_0^{-\infty} S_{II}(E)\langle e^{itH_1}\varphi^-|E^-\rangle\langle^+E|\psi^+\rangle\, dE
\end{aligned}$$

$$= \int^{-\infty}_0 S_{II}(E)\langle \varphi^- | E^- \rangle \langle {}^+E | \psi^+ \rangle \, e^{-itE} \, dE \ . \qquad (4\text{-}67)$$

The later identity is a consequence of (4-66). On the other hand, by hypothesis we have $|S_{II}(E)| \leq |P(E)|$ where P(E) is a polynomial on E. Thus,

$$|S_{II}(E)\langle \varphi^- | E^- \rangle \langle {}^+E | \psi^+ \rangle| \leq |P(E)| \, |\langle \varphi^- E^- \rangle \langle {}^+E | \psi^+ \rangle| \ . \qquad (4\text{-}68)$$

When E runs out R, $\langle \varphi^- | E^- \rangle \langle {}^+E | \psi^+ \rangle$ is a function in S and consequently $P(E)\langle \varphi^- | E^- \rangle \langle {}^+E | \psi^+ \rangle$ belongs to S and hence it is integrable as a function on R. Therefore,

$$\eta(E) = \begin{cases} S_{II}(E) \langle \varphi^- | E^- \rangle \langle {}^+E | \psi^+ \rangle & \text{if } E \in R^- \\ 0 & \text{if } E \in R^+ \end{cases} \qquad (4\text{-}69)$$

is an integrable function of real variable.

Let

$$\varphi(t) = - F \, [\eta(E)] = - \int^{\infty}_{-\infty} \eta(E) e^{-itE} \, dE$$
$$= \int^{-\infty}_0 S_{II}(E) \langle \varphi^- | E^- \rangle \langle {}^+E | \psi^+ \rangle \, e^{-itE} \, dE \ . \qquad (4\text{-}70)$$

To achieve our final conclusion, we need to state the following result[9]:

<u>Theorem</u> (Riemann-Lebesgue). If $\eta(E)$ is an integrable function, its Fourier transform is continuous, has continuous derivatives to all orders and vanishes at the infinite.

From this theorem it is obvious that

$$0 = -\lim \ \varphi(t) \quad \text{as } t \to \infty$$

$$= \lim_{t \to \infty} \int^{-\infty} S_{II}(E) \langle \varphi^- | E^- \rangle \langle {}^+E | \psi^+ \rangle \, e^{-itE} \, dE$$
$$= \lim_{t \to \infty} \langle \varphi^- | (e^{-itH_1})^{\times} | \text{background} \rangle = 0 \qquad (4\text{-}71)$$

or equivalently

$$\lim_{t \to \infty} \ (e^{-itH_1})^{\times} | \text{background} \rangle = 0 \ . \qquad (4\text{-}72)$$

By using similar arguments, we can prove that

$$\lim_{t \to -\infty} \ (e^{-itH_1})^{\times} | \text{background}^* \rangle = 0 \qquad (4\text{-}73)$$

where $|\text{background}^*\rangle \in (\Phi^+)^{\times}$ is the integral term in (4-31-A).

We summarize our above results: The dynamical group e^{-itH_1} splits into two semigroups. For $t < 0$, $(e^{-itH_1})^{\times}$ affects only to the growing part of the resonance and is not defined on the decaying part. When $t > 0$, $(e^{-itH_1})^x$ is the evolution semigroup which gives the time behavior for the decay and it does not affect the growing part of the resonance.

This split leads to the irreversibility of the decay and growth phenomena.

We finish this section with an important remark. The vectors $|Z_R^-\rangle$ and $|Z_R^{*+}\rangle$ are well defined functionals in $(\Phi^-)^x$ and $(\Phi^+)^x$ respectively. One may ask whether they are normalizable vectors in H. The answer is no. To prove it, we first recall that

$$\lim_{t \to \infty} (e^{-itH_1})^x |Z_R^-\rangle = \lim_{t \to \infty} e^{-itE} e^{-t\Gamma/2} |Z_R^-\rangle = 0 \qquad \text{(4-74-A)}$$

and

$$\lim_{t \to \infty} (e^{-itH_1})^x |Z_R^{*+}\rangle = \lim_{t \to \infty} e^{-itE} e^{t\Gamma/2} |Z_R^{*+}\rangle = 0 \ . \qquad \text{(4-74-B)}$$

Proposition V. - Assume that $|Z_R^-\rangle$ were a vector in the Hilbert space H. Then, it would belong to the subspace H_{ac}. The same result is valid for $|Z_R^{*+}\rangle$. (Recall that $H = H_p \oplus H_{ac}$.)

Proof. - First, we need two previous remarks.

a) $|Z_R^-\rangle$ and $|Z_R^{*+}\rangle$ can never be zero, since they are eigenvectors of $H_1{}^x$ with nonzero eigenvalue (see also Appendix VII).

b) if $h \in H_{ac}$; $e^{-itH_1} h = (e^{-itH_1})^x h \in H_{ac}$ whenever the extension exists. (See comments at the beginning of the present section.) Now, we give the proof of this proposition. Assume that $|Z_R^-\rangle \in H$, but $|Z_R^-\rangle \notin H_{ac}$.

α) Either $|Z_R^-\rangle \in H_p$. Then, $(e^{-itH_1})^x |Z_R^-\rangle = 0$ for $t > 0$. To see this, take $f \in \Phi^-$. We have $\langle f|(e^{-itH_1})^x |Z_R^-\rangle = \langle e^{itH_1} f|Z_R^-\rangle = 0, \forall f \in \Phi^- \subset H_{ac}$.

$\Rightarrow (e^{-itH_1})^x |Z_R^-\rangle = 0$. Proposition IV implies that $|Z_R^-\rangle = 0$, which is absurd.

β) Or $|Z_R^-\rangle = |\mu\rangle + |\nu\rangle$ with $|\mu\rangle \in H_p$ and $|\nu\rangle \in H_{ac}$.

$(e^{-itH_1})^x |Z^-_R\rangle = (e^{-itH_1})^x |\mu\rangle + (e^{-itH_1})^x |\nu\rangle = (e^{-itH_1})^x |\nu\rangle = e^{-itH_1} |\nu\rangle \in H_{ac}$.

The second identity comes after α). This result combined with Proposition IV yielded

$$e^{-itH_1} |\nu\rangle = \exp\{-itZ_R\} |Z_R^-\rangle \in H_{ac},$$

and therefore $|Z_R^-\rangle \in H_{ac}$, which contradicts our assumption.

Thus, $|Z_R^-\rangle$ might be in H_{ac} if it were a normalizable vector in H. The same kind of arguments can be used for $|Z_R^{*+}\rangle$.

Proposition VI. - $|Z_R^-\rangle$ and $|Z_R^{*+}\rangle$ cannot belong to H_{ac}.

Proof. - Since $(e^{-itH_1})^x$ $(t > 0)$ is the extension of the operator e^{-itH_1}, as defined on H_{ac}, to $(\Phi^-)^x$, we have

$$(e^{-itH_1})^{\times} |Z_R^-\rangle = e^{-itH_1} |Z_R^-\rangle \in \mathcal{H}_{ac} \,. \tag{4-75}$$

e^{-itH_1} is a unitary operator on $\mathcal{H}_{ac}$ and, therefore, leaves the norm of each vector in $\mathcal{H}_{ac}$ invariant. Thus,

$$\| |Z_R^-\rangle\| = \lim \| |Z_R^-\rangle\| = \lim \|e^{-itH_1} |Z_R^-\rangle\|$$
$$= \| \lim e^{-itH_1} |Z_R^-\rangle\| = \| \lim (e^{-itH_1})^{\times} |Z_R^-\rangle\| = 0 \text{ as } t \to \infty. \tag{4-76}$$

Only the zero vector can have zero norm. Therefore, $|Z_R^-\rangle = 0$, which is absurd. Similar arguments prove the same result for $|Z_R^{*+}\rangle$.

An obvious corollary of Proposition V and VI says that both $|Z_R^-\rangle$ and $|Z_R^{*+}\rangle$ cannot belong to $\mathcal{H}$.

Nevertheless, we can construct a representation in which the growing and decaying Gamow states can be represented by vectors having a finite norm. To this end, let us recall the diagram in Chapter III and write

$$(\theta^{\times})^{-1} U^{\times}[(\Omega^-)^{\times}]^{-1} |Z^-_R\rangle = G_+(Z_R) \in (\Delta_+)^{\times} \tag{4-77-A}$$

$$(\theta^{\times})^{-1} U^{\times}[(\Omega^+)^{\times}]^{-1} |Z_R^{*+}\rangle = G_-(Z^*_R) \in (\Delta_-)^{\times} \,. \tag{4-77-B}$$

According to (3-87), we have

$$G_+(Z_R) = -(2\pi i)^{-1}(E - Z_R)^{-1} \in \mathcal{H}_+ \subset L^2(R) \tag{4-78-A}$$

$$G_-(Z^*_R) = (2\pi i)^{-1}(E - Z_R^*)^{-1} \in \mathcal{H}_- \subset L^2(R) \,. \tag{4-78-B}$$

However, $(E - Z_R)^{-1} \notin S$ and $(E - Z_R^*)^{-1} \notin S$, because

$$\lim_{E \to \pm\infty} E(E - Z_R)^{-1} = \lim_{E \to \pm\infty} E(E - Z_R^*)^{-1} = 1 \,.$$

$G_+(Z_R)$ represents the decaying state in the triplet $\Delta_+ \subset \mathcal{H}_+ \subset (\Delta_+)^{\times}$ and $G_-(Z_R)$ the growing state in $\Delta_- \subset \mathcal{H}_- \subset (\Delta_-)^{\times}$. In addition, we know from (3-95) that

$$\hat{\varepsilon}^{\times} G_+(Z_R) = Z_R\, G_+(Z_R) \tag{4-79-A}$$

$$\hat{\varepsilon}^{\times} G_-(Z^*_R) = Z^*_R\, G_-(Z^*_R) \tag{4-79-B}$$

where the relation between $\hat{\varepsilon}$ and H_l is given by (3-90). The triplets $\Delta_\pm \subset H_\pm \subset \Delta^x_\pm$ are a representation of the triplets $\Phi^{\mp} \subset H_{ac} \subset (\Phi^{\mp})^x$ in a weak sense only, since these spaces cannot be related by means of a unitary operator (see the diagram in Chapter III). Henceforth, we shall call (4-78) the Breit-Wigner representation of $|Z^-_R\rangle$ and $|Z_R{}^{*+}\rangle$. $G_+(Z_R)$ and $G_-(Z^*{}_R)$ are the Gamow vectors in the Breit-Wigner representation. They fulfill the following equations

$$(e^{-it\hat{\varepsilon}})^x\, G_+(Z_R) = \exp\{-itZ_R\}\, G_+(Z_R), \quad t>0 \tag{4-80-A}$$

$$(e^{-it\hat{\varepsilon}})^x\, G_-(Z^*{}_R) = \exp\{-itZ_R{}^*\}\, G_-(Z^*{}_R), \quad t<0 \; . \tag{4-80-B}$$

We leave the proof of these equations to the reader.

What are the advantages of the Breit-Wigner representation? As we shall see in the next section, virtual states are represented by non-normalizable vectors in the Breit-Wigner representation in sharp contrast to the Gamow vectors $|Z_R{}^-\rangle$ and $|Z_R{}^{*+}\rangle$ whose Breit-Wigner representations $G_+(Z_R)$ and $G_-(Z^*{}_R)$ are obviously normalizable. On the other hand, both Gamow vectors and virtual state vectors are non-normalizable if we realize them in any of the other triplet systems shown in the diagram of Chapter III. Thus, the Breit-Wigner representation nicely distinguish between virtual state and resonance phenomena.

Finally, note that (4-77) relates two non-normalizable functionals to two normalizable ones. This fact could be surprising and it is due to the non unitarity of θ as well as of its extension θ^x when it is restricted to $H_\pm$. As a matter of fact, θ^x imbeds $H_\pm$ into $(D_\pm)^x$ (see diagram in Chapter III). However, if, say, $f \in H_+$, $\theta^x f$ may or may not be in $L^2(R^+)$. The Gamow vectors provide an example of members of $H_\pm$ for which their images by θ^x are not in $L^2(R^+)$, for if

$$\left\{ \begin{array}{l} \theta^x\, G_+(Z_R) \\ \theta^x\, G_-(Z_R{}^*) \end{array} \right\}$$

were in $L^2(R^+)$,

$$\left\{ \begin{array}{l} |Z_R{}^-\rangle \\ |Z_R{}^{*+}\rangle \end{array} \right\}$$

would be in H_{ac}, since on $L^2(R^+)$.

$$\left\{\begin{array}{l} (\Omega^-)^x(U^{-1})^x = \Omega^- U^{-1} \\ \\ (\Omega^+)^x(U^{-1})^x = \Omega^+ U^{-1} \end{array}\right\} \quad \text{which are unitary operators} \qquad \begin{array}{r}(4\text{-}81\text{-}A) \\ \\ (4\text{-}81\text{-}B)\end{array}$$

from $L^2(R^+)$ onto H_{ac}. This is the example that we mentioned already at the end of Section III of Chapter III.

III. VIRTUAL STATES

The formalism described in the previous section can also be applied to virtual states. In order to do that the spaces that we constructed need to be slightly modified. Virtual states correspond to simple poles of the S-matrix on the negative real axis in the second sheet. For more on virtual states we refer to References 3, 4, 10-12.

In order to apply the above formalism to virtual states, we need to make sure that our functions in $\Delta_\pm$ are analytic at the value of $E \in R^-$ at which the pole of $S_{II}(\omega)$ appears. To accomplish this, we choose those functions in $\Delta_\pm$ which admit an analytic continuation into a strip beyond the real axis. These classes of functions form a linear space which we shall call $\hat{\Delta}_\pm$.

$\hat{\Delta}_\pm$ will be constructed as follows: Let α be any positive number. $S^{(\alpha)}$ is the linear space of all the infinitely differentiable functions $\varphi(x)$ such that

$$e^{\alpha|x|}\, x^m\, (d^n/dx^n)\varphi(x) \tag{4-85}$$

is bounded for all, m, n = 0,1,2,.... It is an easy exercise to prove that $S^{(\alpha)} \subset S$. We define a topology on $S^{(\alpha)}$ in the following way: A sequence $\varphi_n(x)$ converges to $\varphi(x)$ in $S^{(\alpha)}$ if and only if

$$\left\| e^{\alpha|x|}\, x^n \frac{d^m}{dx^m} (\varphi_k(x) - \varphi(x)) \right\| \to 0 \tag{4-86}$$

for any n, m = 0,1,2,...., and the norm considered in (4-86) is that one on $L^2(R)$. $S^{(\alpha)}$ with this sense of convergence is a complete nuclear space (and also metrizable).

By $S^{(\alpha)}(R^-)$ we will denote the linear space of all the functions in $S^{(\alpha)}$ which are zero in R^+. Now we define $\hat{\Delta}_+$ by:

$$\hat{\Delta}_+ = F\ \{S^{(\alpha)}(R^-)\} \tag{4-87}$$

i.e., the space of all Fourier transforms of the functions in $S^{(\alpha)}(R^-) \subset S$, $\hat{\Delta}_+ \subset S$. Also because $S^{(\alpha)}(R^-) \subset L^2(R^-)$ and because of the Paley Wiener theorem I, $\hat{\Delta}_+ \subset H_+$. Therefore $\hat{\Delta}_+ \subset H_+ \cap S$ and

$$\hat{\Delta}_+ \subset \Delta_+ \ . \tag{4-88}$$

Any $f(E) \in \hat{\Delta}_+$ can be analytically continued into the strip $\{\omega \in C \text{ s.t. } |\mathrm{Im}\,\omega| < \alpha\}$. This assertion immediately follows from Theorem IV of Chapter 2.

Thus, functions in $\hat{\Delta}_+$ have the following properties:

1. They are analytic for Im $\omega > -\alpha$.
2. They lie in H_+.
3. They belong to S.

We can endow $\hat{\Delta}_+$ with a complete nuclear topology: $f_n(E) \rightarrow f(E)$ in $\hat{\Delta}_+$ if and only if $F^{-1}\{f_n(E)\} \rightarrow F^{-1}\{f(E)\}$ in $S^{(\alpha)}$.

Since $\hat{\Delta}_+ \subset H_+$, we have again a bijective correspondence between $\hat{\Delta}_+$ and the linear space of the restrictions to R^+ of the functions in $\hat{\Delta}_+$. We denote this correspondence again by θ_+ and write

$$\hat{D}_+ = \theta\hat{\Delta}_+ \ . \tag{4-89}$$

$\hat{D}_+$ has all the properties 1-6 listed in Chapter III for D_+. Those properties have their corresponding counterparts on $\hat{\Phi}_- = U^{-1}\hat{D}_+$ and on $\hat{\Phi}^- = \Omega^-\hat{\Phi}_-$.

$\hat{D}_-$ and $\hat{\Delta}_-$ are, respectively, the spaces of all complex conjugates of functions in $\hat{D}_+$ and $\hat{\Delta}_+$. All the functions in $\hat{\Delta}_-$ can be analytically continued into the strip $\{\omega: |\text{Im}\ \omega| < \alpha\}$ so that they have the following properties:

1. They are analytic when Im $\omega < \alpha$.
2. They lie in H_-.
3. They belong to S.

Again, we define, $\hat{\Phi}_+ = U^{-1}\hat{D}_-$; $\hat{\Phi}^+ = \Omega^+\hat{\Phi}_+$ and call θ the bijection from $\hat{\Delta}_-$ onto $\hat{D}_-$. The diagram relating all these spaces and their duals is the same as the diagram given in Chapter III. In particular we have the following R.H.S.:

$$\hat{\Delta}_\pm \subset H_\pm \subset (\hat{\Delta}_\pm)^\times \tag{4-90-A}$$

$$\hat{D}_\pm \subset L^2(R^+) \subset (\hat{D}_\pm)^\times \tag{4-90-B}$$

$$\hat{\Phi}_\mp \subset H \subset (\hat{\Phi}_\mp)^\times \tag{4-90-C}$$

$$\hat{\Phi}^\mp \subset H_{ac} \subset (\hat{\Phi}^\mp)^\times \ . \tag{4-90-D}$$

Many of the properties of the old spaces are also properties of these new spaces. In particular $|Z_R^-\rangle$ and $|Z_R^{*+}\rangle$ are defined on Φ^- and Φ^+ exactly as on the old spaces (see (3-64) and after) and they belong to $(\Phi^-)^\times$ and $(\Phi^+)^\times$, respectively.

Now assume $S_{II}(\omega)$ has, in addition to the resonance pole, a simple pole at $\omega = -E_V < 0$

($E_V > 0$). Take $S_{II}(\omega)\,\langle\varphi^-|\omega^-\rangle\langle^+\omega^*|\psi^+\rangle$ and integrate it over the contour shown in figure 8 . Using the residue theorem, we have

$$\int_C S_{II}(\omega)\langle\varphi^-|\omega^-\rangle\langle^+\omega^*|\psi^+\rangle\, d\omega = 2\pi i\,\{\mathrm{Res}(Z_R) + \mathrm{Res}(-E_V)\} \tag{4-91}$$

where $\mathrm{Res}(Z_R)$ and $\mathrm{Res}(-E_V)$ are the residues of the function under integral sign in (4-91) at Z_R and $-E_V$, respectively. We can deform the contour again. The radius of the semicircle goes to infinity and the straight part near the cut can be pulled toward it. The integral over the infinite semicircle vanishes because of (4-7); so we have in the limit

$$\begin{aligned}
&-\int_0^{\infty} S(E + i0)\langle\varphi^-|E^-\rangle\langle^+E|\psi^+\rangle\, dE \\
&+\int_0^{-E+\varepsilon} S_{II}(E)\,\langle\varphi^-|E^-\rangle\langle^+E|\psi^+\rangle\, dE \\
&+\int_C S_{II}(\omega)\,\langle\varphi^-|\omega^-\rangle\langle^+\omega^*|\psi^+\rangle\, d\omega \\
&+\int_{-E\,-\varepsilon}^{-\infty} S_{II}(E)\,\langle\varphi^-|E^-\rangle\langle^+E|\psi^+\rangle\, dE \\
&\qquad = 2\pi i\,\{\mathrm{Res}(Z_R) + \mathrm{Res}(-E_V)\}
\end{aligned} \tag{4-92}$$

where C_ε is a semicircle of radius ε around $-E_V$ as shown in figure 8.

When $\varepsilon \to 0$, a proper use of complex variables shows that the last three integrals in (4-92) give

$$P\int^{-\infty}_0 S_{II}(E)\,\langle\varphi^-|E^-\rangle\langle^+E|\psi^+\rangle\, dE - \pi i\,\mathrm{Res}(-E_V)\ . \tag{4-93}$$

Here P means principal value. According to (4-92) and (4-93) we have

$$\begin{aligned}
&\int^{\infty}_0 S(E + i0)\,\langle\varphi^-|E^-\rangle\langle^+E|\psi^+\rangle\, dE \\
&= P\int^{-\infty}_0 S_{II}(E)\,\langle\varphi^-|E^-\rangle\langle^+E|\psi^+\rangle\, dE \\
&-2\pi i\,\mathrm{Res}(Z_R) - 3\pi i\,\mathrm{Res}(-E_V)\ .
\end{aligned} \tag{4-94}$$

If s_- and s_V are the residui of $S_{II}(\omega)$ at Z_R and $-E_V$ respectively, we have

$$\begin{aligned}
(\psi^-,\psi^+) = {} & P\int^{-\infty}_0 S_{II}(E)\,\langle\varphi^-|E^-\rangle\langle^+E|\psi^+\rangle\, dE \\
& - 2\pi i s_{-1}\,\langle\varphi^-|Z_R^-\rangle\langle^+Z^*_R|\psi^+\rangle \\
& - 3\pi i s_V\,\langle\varphi^-|-E_V^-\rangle\langle(-E_V)^+|\psi^+\rangle\ .
\end{aligned} \tag{4-95-A}$$

By omitting the arbitrary $\varphi^- \in \Phi^-$, we have formally

$$\psi^+ = |\mathrm{background}\rangle - 2\pi i s_{-1}|Z_R^-\rangle\langle^+Z^*_R|\psi^+\rangle$$

$$- 3\pi i s_V \, |(-E_V)^-\rangle\langle^+ Z^*{}_R|\psi^+\rangle \ . \qquad (4\text{–}96\text{–}A)$$

Here

$$|\text{background}\rangle = P \int^{-\infty}_0 S_{II}(E)|E^-\rangle\langle^+ E|\psi^+\rangle \, dE \ . \qquad (4\text{-}97\text{-}A)$$

ψ^+ is again the decaying state. It is a sum of an exponentially decaying state, plus the background, plus the virtual state vector, $|(-E_V)^-\rangle$ multiplied by a constant.

If we now take the contour of figure 9 we arrive by similar considerations at:

$$\begin{aligned}(\psi^+,\varphi^-) = {} & P \int^{\infty}_0 S_{II}(E) \, \langle\psi^+|E^+\rangle\langle E^-|\varphi^-\rangle \, dE \\ & + 2\pi i \sigma_{-1} \, \langle\psi^+|Z_R{}^{*+}\rangle\langle Z_R{}^-|\varphi^-\rangle \\ & + 3\pi i \sigma_v \, \langle\psi^+|(-E_V)^+\rangle\langle(-E_V)^-|\varphi^-\rangle \ . \end{aligned} \qquad (4\text{-}95\text{-}B)$$

Again, by omitting the arbitrary $\psi^+ \in \Phi^+$ we have the following formal identity in $(\Phi^+)^x$:

$$\begin{aligned}\varphi^- = {} & |\text{background}^*\rangle + 2\pi i s_{-1}|Z_R{}^{*+}\rangle\langle Z_R{}^-|\varphi^-\rangle \\ & + 3\pi i \sigma_v|(-E_V)^+\rangle\langle(-E_V)^-|\varphi^-\rangle \end{aligned} \qquad (4\text{–}96\text{–}B)$$

where

$$|\text{background}^*\rangle = P \int_0^{\infty} \, S_{II}(E)|E^+\rangle\langle E^-|\varphi^-\rangle \, dE \ . \qquad (4\text{-}97\text{-}B)$$

φ^- is the growing state. It is a linear combination of the Gamow growing state, the background, and a new virtual state vector, namely $|(-E_V)^+\rangle$.

Finally, we are going to make a study on the time evolution of the virtual state vectors $|(-E_V)^-\rangle$ and $|(-E_V)^+\rangle$.

<u>Proposition</u>.
For $t > 0$ $\; e^{itH_1} \hat{\Phi}^- \subset \hat{\Phi}^-$ and e^{itH_1} is continuous on $\hat{\Phi}^-$.

For $t < 0$ $\; e^{itH_1} \hat{\Phi}^+ \subset \hat{\Phi}^+$ and e^{itH_1} is continuous on $\hat{\Phi}^+$.

<u>Idea of the proof.</u> Let $t > 0$. We have to prove $e^{itE}\hat{\Delta}_+ \subset \hat{\Delta}_+$. For it, we use Theorem IV in Chapter II. Let $f(E) \in \hat{\Delta}_+$, we need showing $e^{itE}f(E) \in \hat{\Delta}_+$. Since $\hat{\Delta}_+ \subset \Delta_+ \Rightarrow e^{itE}f(E) \in \Delta_+$. It only reemains showing that

$$\int^{\infty}_{-\infty} |e^{it(E+iy)} \, f(E + iy)|^2 \, dE \qquad (4\text{-}98)$$

is uniformly bounded for $|y| \leq b$, $b < \alpha$ and t fixed

$$(4\text{-}98) = \int_{-\infty}^{\infty} |e^{-ty}|f(E + iy)|^2\, dE$$
$$= e^{-2ty} \int_{-\infty}^{\infty} |f(E + iy)|^2\, dE$$
$$\leq e^{2tb}\, k \quad \text{whenever } |y| \leq b \tag{4-99}$$

$f(E) \in \hat{\Delta}_+$ and therefore for $|y| \leq b$ there exists a constant k such that the integral in the third member of (4-99) is bounded by k. Then, use Theorem IV in Chapter II to conclude that $e^{itE}f(E) \in \hat{\Delta}_+$.

Let now $f_n(E) \to f(E)$. If $\varphi_n(t) = \mathcal{F}^{-1} f_n(E)$ and $\varphi(t) = \mathcal{F}^{-1} f(E)$, $\varphi_n(x) \to \varphi(x)$ in the sense of (4-85) $\mathcal{F}^{-1} f_n(E)e^{itE} = \varphi_n(x + t) \to \varphi(x + t) = \mathcal{F}^{-1} f(E)e^{itE}$ which means that in $\hat{\Delta}_+$:

$$e^{itE} f_n(E) \to e^{itE} f(E) \qquad t > 0\,. \tag{4-100}$$

From this and the continuity of θ, U^{-1} and Ω^-, the proposition follows.

Let now $\varphi^- \in \hat{\Phi}^-$ and $f_+(E) = \theta^{-1}\, U(\Omega^-)^{-1}\, \varphi^- \in \hat{\Delta}_+$; $t > 0$. We then obtain:

$$\langle \varphi^- | (e^{-itH_1})^x | (-E_V)^- \rangle = \langle e^{itH_1}\, \varphi^- | (-E_V)^- \rangle = [e^{-itE_V}\, f_+(-E_V)]^* = e^{itE_V}\, [f_+(-E_V)]^*$$
$$= e^{itE_V}\, \langle \varphi^- | (-E_V)^* \rangle \tag{4-101-A}$$

which implies

$$(e^{-itH_1})^x | (-E_V)^- \rangle = e^{itE_V}\, | (-E_V)^- \rangle\,; \qquad t > 0\,. \tag{4-102-A}$$

Similarly we obtain if $\psi^+ \in \hat{\Phi}^+$ and $g_-(E) = \theta^{-1}\, U(\Omega^+)^{-1}\, \psi^+ \in \hat{\Delta}_-$

$$\langle {}^+\psi | (e^{-itH_1})^x | (-E_V)^+ \rangle = \langle e^{itH_1} \psi^+ | (-E_V)^+ \rangle$$
$$= [e^{-itE_V}\, g_-(-E_V)]^* = e^{itE_V}\, [g_-(-E_V)]^*$$
$$= e^{itE_V}\, \langle \psi^+ | (-E_V)^+ \rangle \qquad t < 0\,. \tag{4-101-B}$$

Hence

$$(e^{-itH_1})^x | (-E_V)^+ \rangle = e^{itE_V}\, | (-E_V)^+ \rangle \qquad t < 0\,. \tag{4-102-B}$$

Either of the vectors $|(-E_V)^-\rangle$ or $|(-E_V)^+\rangle$ can be used to describe the same virtual state of $-E_V$. From (4-102), we see that these two vectors describe states that do not decay.

The feature which distinguishes the virtual state vectors from the Gamow vectors becomes apparent when we use the Breit-Wigner representation, i.e., describe them by using the triplets $(\hat{\Delta}_\pm)^x \supset \mathcal{H}_\pm \supset \hat{\Delta}_\pm$. The vector in $(\hat{\Delta}_+)^x$ defined by

$$\delta_-(-E_V) \equiv (\theta^x)^{-1}\, U^x [(\Omega^-)^x]^{-1} | (-E_V)^- \rangle \tag{4-103-A}$$

and the vector in $(\hat{\Delta}_-)^x$ defined by

$$\delta_+(-E_V) \equiv (\theta^x)^{-1} U^x[(\Omega^-)^x]^{-1}|(-E_V)^+\rangle \qquad (4\text{-}103\text{-}B)$$

are nothing other than the Dirac delta at the point $-E_V$ defined on $\hat{\Delta}_+$ and $\hat{\Delta}_-$ respectively. (Actually, they are the complex conjugate of the Dirac delta, i.e., the functional assigning to every function in $\hat{\Delta}_\pm$ the complex conjugate of its value at $-E_V$.) We know that a Dirac delta type functional can never be a square integrable function and therefore be in $H_\pm$ or in $L^2(R)$. This distinguishes virtual state vectors from Gamow vectors which belong either to H_+ or to H_-. In conclusion, in the Breit-Wigner representation the virtual state vectors are given by

$$(\hat{\Delta}_-)^x \ni \delta_-(-E_V) \notin L^2(R) \qquad (4\text{-}104\text{-}A)$$

$$(\hat{\Delta}_+)^x \ni \delta_+(-E_V) \notin L^2(R) \ . \qquad (4\text{-}104\text{-}B)$$

As a summary of our results we present the following scheme which shows the correspondence between vector description of the physical states and the properties of the S-matrix.

	S-matrix description	Vector space description
Stationary state	**Pole in the negative real axis**	**Eigenvector of H**
Scattering state	**Cut along the spectrum of H_I**	**Generalized eigenvector of H (Dirac ket)**
Quasistationary state	**Pole in the second sheet**	**Gamow vector**
Resonance	**Pair of poles in the second sheet**	**Pair of Gamow vectors**
Virtual state	**Pole in the negative real axis in the second** sheet	**Pair of virtual state vectors**

References for Chapter IV

1.- A. Bohm, J. Math. Phys. 21, 1040, (1980); J. Math. Phys 22 , 2813 (1981); Physica 124A , 103 (1984).- E.C.G. Sudarshan, C.B. Chiu, V. Gorini, Phys. Rev. D 18, 2914, (1978).

2.- For a classification of the spectrum, see Reference 29 in Chapter I.

3.- A Bohm,"Quantum Mechanics: Foundations and Applications" 2nd edition. Springer Verlag. Berlin and New York (1986).

4.- H.M. Nussenzveig, "Causality and Dispersion Relations", Academic Press, London (1972).

5.- A. Bohm, J. Math. Phys. 22, 2813 (1981). Also Reference 7 in Chapter II.

6.- See Reference 7 on Chapter III.

7.- See for instance, J.E. Marsden, "Basic Complex Analysis", Freemann, San Francisco (1973).

8.- L. Fonda, G.C. Ghirardi and A. Ri
mini, Rep. Prog. Phys., 41, 578 (1978).

9.- See Reference 20 in Chapter II.

10.- J.M. Blatt and V.F. Weisskopf, "Theoretical Nuclear Physiscs", Springer Verlag, New York (1979).

11.- M. Gadella, J. Math. Phys. 24, 2142 (1983)

12.- J.R. Taylor, "Scattering Theory", Wiley, New York (1972).

APPENDIX I

LINEAR SPACES. LINEAR OPERATORS

A linear scalar product space Φ is (not necessarily a space of functions) a set of mathematical (imagined) objects for which three operations are defined.

I. Addition of elements of Φ, which has the following properties:

(i) For h, f, $g \in \Phi$, $f + g \in \Phi$ and

$$f + g = g + f$$

$$(f + g) + h = f + (g + h) = f + g + h.$$

(ii) There existsan element $0 \in \Phi$ such that $f + 0 = f$

(iii) $h + f = g$ has a unique solution:

$$h = g - f.$$

II. Multiplication by complex numbers, which has the following properties:

$\alpha(f + g) = \alpha f + \alpha g$, $\alpha \in C$ = set of complex numbers

$(\alpha + \beta)f = \alpha f + \beta f$, $\alpha, \beta \in C$

$\alpha(\beta f) = (\alpha\beta)f$; $0f = 0 \in \Phi$

III. Scalar product of two elements $\varphi, \psi \in \Phi$,i.e., with every pair of elements $\varphi, \psi \in \Phi$ there is associated a complex number (φ, ψ) with the following properties:

(i) $(\varphi, \psi) = (\psi, \varphi)^*$, where $(\psi, \varphi)^*$ is the complex conjugate of (φ,ψ).

(ii) $(\varphi, \alpha_1\psi_1 + \alpha_2\psi_2) = \alpha_1(\varphi, \psi_1) + \alpha_2(\varphi, \psi_2)$.

(iii) $(\varphi, \varphi) \geq 0$ with $(\varphi, \varphi) = 0$ only for $\varphi = 0$

As a consequence of (i) and (ii), we can obtain $(\alpha_1\varphi_1 + \alpha_2\varphi_2, \psi) = \alpha_1^* (\varphi_1, \psi) + \alpha_2^* (\varphi_1, \psi)$.

A linear operator A is (not necessarily a differential operation) a map from a subspace D(A) of Φ into Φ which has the property:

$$A(\alpha_1 f_1 + \alpha_2 f_2) = \alpha_1 A(f_1) + \alpha_2 A(f_2), \quad \alpha_1, \alpha_2 \in C, f_1, f_2 \in \Phi$$

D(A) is the domain of the operator A. We say that A is bounded when we have a positive constant k such that

$$||A\psi|| \leq k||\psi||, \quad \forall\psi \in D(A).$$

It can be proved that a linear operator A is bounded if and only if it is continuous with respect to the topology given by the following metric:

$$d(f, g) = ||f - g|| = \{(f - g, f - g)\}^{1/2}.$$

If A is bounded, its domain is always Φ.

APPENDIX II
ALGEBRA

A set $\mathcal{A}$ is an associative algebra with unit element iff

(i) $\mathcal{A}$ is a linear space over C.

(ii) For every pair A, B $\in \mathcal{A}$, there is defined a product

A,B $\in \mathcal{A}$ such that

$(AB)C = A(BC)$

$A(B + C) = AB + AC.$

$(A + B)C = AC + BC.$

$(\alpha A)B = A(\alpha B) = \alpha AB.$

(iii) There exitst an element $1 \in \mathcal{A}$ such that $1A = A$, for all $A \in \mathcal{A}$.

A set K of elements of $\mathcal{A}$ is called a system of generators iff the smallest closed subalgebra with unit which contains K coincides with $\mathcal{A}$. The element 1 is not to be called a generator.

Let the elements of K be called X_i $i = 1,2,\ldots,n$. Then each element of $\mathcal{A}$ can be written

$$A = C^0 + \Sigma\, C^i X_i + \Sigma\, C^{ij} X_i X_j + \ldots.$$
$$C^0, C^i, \ldots, \in C \qquad \text{(II-1)}$$

We restrict ourselves to the case that n is finite and we will not discuss topologies on $\mathcal{A}$, i.e., we assume that the above sums for every A are arbitrarily large but finite. Defining algebraic relations are relations among the generators

$$P(X_i) = O \qquad \text{(II-2)}$$

where $P(x_i)$ is a polynomial with complex coefficients of n variables x_i. An element $B \in \mathcal{A}$

$$B = b^0 + \Sigma\, b^i X_i + \Sigma\, b^{ij} X_i X_j + \ldots ,$$
$$b^i \ldots \in C \qquad \text{(II-3)}$$

is equal to the element A iff (II-3) can be brought inti the same form (II-1) with the same coefficients $C^0, C^i, C^{ij}, \ldots$ by the use of the defining relations (II-2).

The best known examples of associative algebras are the enveloping algebras $\varepsilon(G)$ of Lie groups G.

An enveloping algebra is the associative algebra generated by $X_1, X_2, \ldots, X_n$ in which the multiplication is defined by the commutation relations

$$P(X_i) = X_i X_j - X_j X_i - \Sigma\, C_{ij}{}^k X_k = 0 \qquad \text{(II-4)}$$

$C_{ij}{}^k$ are called the structure constants of the Lie groupG. The commutation relations (II-4) do in general not suffice to fully define an enveloping algebra of linear operators in linear topological spaces that fulfill a given commutation relation. In order to specify a particular algebra of operators one has to require additional algebraic relations and other properties (such as $\Sigma X^2{}_i$ be e.s.a.)

APPENDIX III
LINEAR OPERATORS ON HILBERT SPACES

We here recall a few of the definitions and important results from the theory of linear operators on Hilbert spaces.

Let $A: D(A) \to H$ be a linear but not necessarily bounded operator on a Hilbert space H and let $D(A)$ be dense in H. We define,

$D(A^+) = \{f \in H$ such that there exists $g \in H$ with $(f, Ah) = (g, h), \ \forall h \in D(A)\}$

This vector g is unique. If we define $g = A^+f$, A^+ is a linear operator with domain $D(A^+)$ and we obviously have

$(f, Ah) = (A^+f, h)$.

A^+ is called the ajoint of A.

In the sequel we always assume that D(A) is dense in H (any vector in H is a limit of a Cauchy sequence in D(A)). An operator A is called symmetric if $A \subset A^+$ (i.e., $D(A) \subset D(A^+)$) and $Af = A^+f$ for every $f \in D(A)$. It is called self-adjoint if $A = A^+$.

An operator A on H is called closed if the relations

$\lim f_n = f$; $\lim Af_n = g$; $f_n \in D(A)$.

imply $f \in D(A)$ and $Af = g$. Closedness is a weaker condition than continuity since, if an operator A on H is continuous, then

$\lim f_n = f$ for $f_n \in D(A)$

implies that the sequence $\{Af_n\}$ converges, While if it were closed only, then the convergence of the sequence $\{f_n\}$ for $f_n \in D(A)$ would not imply the convergence of the sequence $\{Af_n\}$. However, if A is closed then, in particular, it has the property that two sequences $\{Af_n\}$ and $\{Ag_n\}$ cannot converge to different limits if the corresponding sequences $\{f_n\}$ and $\{g_n\}$ converge to the same limit. A^+ is always closed.

If A is not closed, it is sometimes possible to find an extension of A which is closed. An operator A admits a closure A iff the relations

$f_n \in D(A)$; $h_n \in D(A)$; $\lim f_n = f$; $\lim h_n = f$

$\lim A f_n = g$; $\lim A h_n = g'$

imply $g = g'$. In this case D(A) consists of all $f \in H$ for which there exists a sequence $\{f_n\} \in D(A)$, which satisfies the conditions:

(i) $\lim f_n = f$

(ii) $\{Af_n\}$ converges.

Then by definition

$Af = \lim Af_n$.

An operator A on $D(A) \subset H$ is called essentially self-adjoint (e.s.a.) if A is self-adjoint. Physical observables are assumed to be represented by essentially self-adjoints operators.

An operator A on H is called Hermitian if A is bounded and self-adjoint. An operator U on H is called unitary if $\|U\varphi\| = \|\varphi\|$, $\forall\ \varphi \in H$, and it is onto. Unitary operators satisfies the relations $U^+U = UU^+ = I$, the identity on H. Equivalently, $U^+ = U^{-1}$. A unitary operator is one-to-one, onto and continuous mapping. If we remove the condition of being onto, the operator is called isometric.

Let H and G be two Hilbert spaces with respective norms $\|.\|_H$ and $\|.\|_G$. A linear mapping U from H into G is unitary if it is onto and $\|Uf\|_G = \|f\|_H$ for all f in H. Note that this implies that U is one-to-one. Therefore U has an inverse $U^{-1} : G \rightarrow H$ which is also unitary.

The range (i.e., its image) of an isometric operator U on H is a Hilbert space K (closed subspace of H). U is unitary from H onto K .

If U is unitary or isometric $(f, g) = (Uf, Ug)$, for all $f, g \in H$.

Let M be a closed subspace of H and $M^{\perp}$ the closed subspace of all the vectors which are orthogonal to each vector of M . Then, $H = M \oplus M^{\perp}$ in the sense thst every $f \in H$ can be written uniquely as $f = h + g$, with $h \in M$ and $g \in M^{\perp}$. An operator P which associates wich each $f \in H$ its projection h in M is called a projection operator on H. Projection operators are linear, Hermitian and idempotent (i.e. $P^2 = P$); and every linear, Hermitian, idempotent operator is a projection operator. Two projection operators are called orthogonal iff $P_1P_2 = 0$; in this case P_1H and P_2H are orthogonal subspaces of H, i.e., for every $h_1 \in P_1H$ and every $h_2 \in P_2H$, we have $(h_1, h_2) = 0$.

APPENDIX IV

PROOF OF THE LEMMA (1-41) CHAPTER I BY INDUCTION

From the commutation relation follows:

$$(N + I)a^+ = a^+(N + I)$$

$$a(N + I) = (N + I)a \qquad \text{(IV-1)}$$

(1-41) is true for $p = 1$, because

$$(\varphi, a(N + I)a^+\varphi) = (\varphi, aa^+(N + I)\varphi) = (\varphi, (N + I)(N + 2I)\varphi)$$

$$=(\varphi,(N + I)^2 \varphi) + (\varphi, (N + I)\varphi) = (\varphi, (N + I)^2\varphi)$$

because $(\varphi, (N + I)\varphi) \leq (\varphi, (N + I)^2\varphi)$.

The latter inequality comes from the fact that, since N has a nonnegative spectrum, it has to be positive, i. e. , $(\varphi, N\varphi) \geq 0$ for all $\varphi \in D(N)$. Then,

$$(\varphi, (N + I)^2\varphi) - (\varphi, (N + I)\varphi)$$

$$= (\varphi, N^2\varphi) + 2(\varphi, N\varphi) + (\varphi, \varphi) - (\varphi, N\varphi) - (\varphi, \varphi)$$

$$= (N\varphi, N\varphi) + (\varphi, N\varphi) = ||N\varphi||^2 + (\varphi, N\varphi) \geq 0.$$

Assume (1-41) is true for $p = q$; i.e.

$$(\varphi, a(N + I)^q a^+\varphi) \leq k(\varphi,(N + I)^{q+1}\varphi) \quad \forall \varphi \qquad \text{(IV-2)}$$

and calculate

$$(\varphi, a(N + I)^{q+1} a^+\varphi) = (\varphi,a(N + I)(N + I)^{q-1}(N + I) a^+\varphi) =$$

because of (IV-1)

$$= (\varphi, (N + 2I)a(N + I)^{q-1} a^+(N + 2I)\varphi) \leq k ((N + 2I)\varphi, (N + I)^q(N + 2I)\varphi) =$$

because (IV-2) is valid for every φ, in particular for $\psi = (N + 2I)\varphi$

$$=k[((N + I)\varphi, (N + I)^q(N + I)\varphi) + (\varphi,(N + I)^q\varphi)$$

$$+ ((N + I)\varphi, (N + I)^q\varphi) + (\varphi, (N + I)^q(N + I)\varphi)]$$

$$= k [(\varphi, (N + I)^{q+2}\varphi) + (\varphi, (N + I)^q\varphi)$$

$$+ (\varphi, (N + I)^{q+1}\varphi) + (\varphi, (N + I)^{q+1}\varphi)].$$

If r and s are positive integers and $r \leq s$, $(\varphi, (N + I)^r\varphi) \leq (\varphi, (N + I)^s\varphi)$ for all φ. This is a consequence of the spectral properties of N (see also (1-30)). Thus, the latter expresion is less or equal to

$$4k(\varphi, (N + I)^{q+2}\varphi).$$

Consequently, (1-41) has been shown to be also fulfilled for $p = q + 1$ and, therefore, it is true for any integer p.

APPENDIX V

We will show in this Appendix that the defining assumptions, i.e., (1-1), (1-2), (1-3) and the τ_Φ-continuity of P and Q, will lead to the Schroedinger representation on the Schwartz space S(R).

We choose one of the energy eigenstates φ_n , n = 1,2,... and calculate

(using $a\varphi_n = \sqrt{n}\ \varphi_{n-1}$, $a^+ \varphi_n = \sqrt{n+1}\ \varphi_{n+1}$):

$$x\langle\varphi_n|x\rangle = \langle Q\varphi_n|x\rangle = \frac{\hbar}{2m\omega}\left(\sqrt{n}\ \langle\varphi_{n-1}|x\rangle + \sqrt{n+1}\ \langle\varphi_{n+1}|x\rangle\right) \tag{V-1}$$

for n = 0 one obtains:

$$\sqrt{2}\ \frac{x}{\sqrt{\hbar/m\omega}}\ \langle\varphi_0|x\rangle = \sqrt{1}\ \langle\varphi_1|x\rangle \tag{V-2}$$

(V-1) is a recurrence relation for $\langle\varphi_n|x\rangle$, which can be brought into a well known form by introducing

$$y = \frac{x}{\sqrt{\hbar/m\omega}}\ ,\quad f_n(y) = \sqrt{2^n n!}\ \frac{\langle\varphi_n|x\rangle}{\langle\varphi_0|x\rangle}$$

(assuming $\langle\varphi_0|x\rangle \neq 0$) (V-3)

(V-1) is then written

$$f_{n+1}(y) = 2yf_n(y) - 2nf_{n-1}(y);\quad n = 1,2,... \tag{V-4}$$

$$f_1(y) = 2yf_0(y);\quad f_0(y) = 1.$$

(V-4) are the well known recurrence relations for the Hermite polynomials and have solutions for any complex y.

Thus, we conclude that for any complex value x, there is an antilinear (not neccesarily continuous) functional $|x\rangle = F_x$ which is a generalized eigenvector.

The functionals $\langle\varphi_n|x\rangle$ are given by

$$\langle\varphi_n|x\rangle = \frac{1}{\sqrt{2^n n!}}\ \langle\varphi_0|x\rangle\ H_n\left(\frac{x}{\sqrt{\hbar/m\omega}}\right);\quad x \in C \tag{V-5}$$

As the operator Q is e.s.a. (as a consequence of (1-3)), the spectrum of Q must be real[1]. Thus the spectrum is only a suset of generalized eigenvalues and in the spectral decomposition

$$\varphi_n = \int d\mu(x)\ |x\rangle\langle x|\varphi_n\rangle \tag{V-6}$$

only those generalized eigenvectors and those functionals $\langle x|\varphi_n\rangle = \langle\varphi|x\rangle^*$ appear for which x is real. To find the spectrum of Q then means to determine which real values appear in (V-6).

If we consider φ_n (or any $\varphi \in \Phi$) as a functional at the generalized eigenvector $F_{x'} = |x'\rangle \in \Phi^x$, $x' \in$ spectrum of Q, then, according to (1-67), we obtain from (V-6):

$$\varphi_n(F_{x'}) = \langle x'|\varphi_n\rangle = \int d\mu(x)\ \langle x'|x\rangle\langle x|\varphi_n\rangle \tag{V-7}$$

Thus, $d\mu(x)\langle x'|x\rangle$ must be the Dirac measure, i.e. the distribution $\langle x'|x\rangle$ defined by (V-7) must have the property of the Dirac δ distribution:

$$d\mu(x)\langle x'|x\rangle = dx\ \delta(x' - x) \tag{V-8}$$

We consider now the generalized eigenvectors of the operator P:

$$P^x|p\rangle = P^xF_p = pF_p \tag{V-9}$$

or $\langle P\varphi|p\rangle = p\langle\varphi|p\rangle$ for every $\varphi \in \Phi$.

Using $P = -i(\hbar m\omega/2)^{-1/2}(a - a^+)$ one can proceed in complete analogy to the procedure for Q following equation (V-1). One obtains that the set of generalized eigenvalues of P is the complex plane and

$$\langle\varphi_n|p\rangle = i^n \frac{1}{\sqrt{2^n n!}} \langle\varphi_0|p\rangle H_n\left(\frac{p}{\sqrt{\hbar m\omega}}\right). \tag{V-10}$$

In analogy to the case for Q, also in the spectral decomposition of φ_n with respect to the operator P:

$$\varphi_n = \int d\mu(p)\ |p\rangle\langle p|\varphi_n\rangle \tag{V-11}$$

only those $\langle p|\varphi_n\rangle = \langle\varphi_n|p\rangle$ of (V-10) occur for which p is real.

If we consider $\varphi_n \in \Phi$ as the functional on the space Φ^x at the element $F_{p'} \in \Phi^x$, then

$$\langle p'|\varphi_n\rangle = \int d\mu(p)\langle p'|p\rangle\langle p|\varphi\rangle \tag{V-12}$$

so that we conclude

$$d\mu(p)\ \langle p'|p\rangle = dp\ \delta(p' - p) \tag{V-13}$$

The analogy between the p- and x-spectral decomposition as expressed, e.g., by the analogy between (V-5) and (V-10) should have been expected, since assumptions (1-1), (1-2), and (1-3) are symmetric in P and Q.

We now calculate the scalar product of φ_n and φ_n using (V-6) with (V-5) and (V-11) with (V-10).

$$\delta_{n,m} = (\varphi_n, \varphi_m) = \int_{\text{spectrum } Q} d\mu(x)\ \langle\varphi_n|x\rangle\ \langle x|\varphi_n\rangle$$

$$= \frac{1}{\sqrt{2^m m!}} \frac{1}{\sqrt{2^n n!}} \int_{\text{spectrum } Q} d\mu(x) \, |\langle\varphi_0 | x\rangle|^2 \, |H_n\{\frac{x}{\sqrt{\hbar/m\omega}}\}|^2$$

$$= \int_{\text{spectrum } P} d\mu(p) \, \langle\varphi_n | p\rangle\langle p | \varphi_m\rangle$$

$$= \frac{1}{\sqrt{2^m m!}} \frac{1}{\sqrt{2^n n!}} \int_{\text{spectrum } P} d\mu(p) \, |\langle\varphi_0 | p\rangle|^2 \, |H_n\{\frac{p}{\sqrt{\hbar/m\omega}}\}|^2$$

(V-14)

Comparing this with the orthogonality relations for the Hermite polynomials

$$\frac{1}{n! \, 2^n \sqrt{\pi}} \int_{-\infty}^{\infty} dy \, e^{-y} \, H_m(y) \, H_n(y) = \delta_{m,n}$$

(V-15)

and taking into account that the Hermite polynomials are only orthogonal polynomials if associated with the interval $-\infty < y < \infty$ and the weight $\exp\{-y^2\}$ (one can define $H_n(y)$ by (V-15) and derive (V-4) for realy) we conclude:

$$d\mu(x) \, |\langle\varphi_0 | x\rangle|^2 = \sqrt{\frac{m\omega}{\hbar \pi}} \, \exp\{-\frac{x^2 m \omega}{2\hbar}\} dx$$

(V-16)

spectrum $Q = \{x \, / -\infty < x < \infty\}$ (V-17)

and

$$d\mu(p) \, |\langle\varphi_0 | p\rangle|^2 = \sqrt{\frac{1}{m\omega\hbar\pi}} \, \exp\{-\frac{p}{2m\omega\hbar}\} \, dx$$

(V-18)

spectrum $P = \{p \, / -\infty < p < \infty\}$ (V-19)

If we agree to normalize the generalized eigenvectors such that

$\langle x' | x\rangle = \delta(x' - x)$ (V-20)

$\langle p' | p\rangle = \delta(p' - p)$ (V-21)

then according to (V-8) and (V-13)

$d\mu(x) = dx$ and $d\mu(p) = dp$ (V-22)

and

$$\langle \varphi_0 | p\rangle = \frac{1}{\sqrt{2^n n! \sqrt{\pi}}} \sqrt[4]{\frac{m\omega}{\hbar}} H_n\{\frac{x}{\sqrt{\hbar/m\omega}}\} \exp\{-\frac{m\omega}{2\hbar} x^2\}$$

(V-23)

$$\langle \varphi_n | p\rangle = i^n \frac{1}{\sqrt{2^n n! \sqrt{\pi}}} \sqrt[4]{\frac{1}{m\omega\hbar}} H_n\{\frac{p}{\sqrt{\hbar m\omega}}\} \exp\{-\frac{p^2}{2m\omega\hbar}\}$$

(V-24)

Thus far, we know the matrix elements of Q in the basis of generalized eigenvectors of Q

$$\langle x | Q\varphi\rangle = \langle Q\varphi | x\rangle^* = \langle \varphi | Q^x | x\rangle^* = x\langle x | \varphi\rangle \qquad \text{(V-25)}$$

and the matrix elements of P in the basis of generalized eigenvectors of P

$$\langle p | P\varphi\rangle = p\langle p | \varphi\rangle \qquad \text{(V-26)}$$

We now want to calculate the matrix elements of P in the basis of generalized eigenvectors of Q and the matrix elements of Q in the basis of generalized eigenvectors of P. In order to do this, we shall consider φ_n as a functional at the generalized eigenvector $F_p \in \Phi^x$, $p \in$ {spectrum P}, and use the spectral decomposition (V-6):

$$\langle p | \varphi_n\rangle = \int dx \langle p | x\rangle\langle x | \varphi_n\rangle \qquad \text{(V-27)}$$

and then as a functional at the generalized eigenvector $F_x \in \Phi^x$, $x \in$ {spectrum Q}, and use the spectral decomposition (V-11):

$$\langle x | \varphi_n\rangle = \int dp \langle x | p\rangle \langle p | \varphi_n\rangle \qquad \text{(V-28)}$$

$\langle p | \varphi_n\rangle$ and $\langle x | \varphi_n\rangle$ in (V-27), (V-28) are given by (V-23) and (V-24) respectively.

The Hermite polynomials have the following property:

$$i^n e^{-\eta^2/2} H_n(\eta) = \int_{-\infty}^{\infty} d\xi \frac{e^{i\xi\eta}}{\sqrt{2\pi}} e^{-\xi^2/2} H_n(\xi)$$

(V-29)

Inserting (V-23) and (V-24) into this relation follows that

$$\langle \varphi_n | p\rangle = \int_{-\infty}^{\infty} dx \frac{1}{\sqrt{2\pi\hbar}} \exp\{ixp/h\}\langle \varphi_n | x\rangle$$

(V-30)

or taking the complex conjugate

$$\langle p|\varphi_n\rangle = \int_{-\infty}^{\infty} dx \frac{1}{\sqrt{2\pi\hbar}} \exp\{-ixp/\hbar\}\langle x|\varphi_n\rangle$$

(V-31)

Comparing (V-31) with (V-27) we find that the distributions $\langle p|x\rangle$ are given by

$$\langle p|x\rangle = (2\pi\hbar)^{-1/2} e^{-i(xp/\hbar)} \quad \text{(V-32)}$$

In the same way one obtains from (V-29) and (v-28) [see (V-23) and (V-24)]

$$\langle x|p\rangle = (2\pi\hbar)^{-1/2} e^{i(xp/\hbar)} \quad \text{(V-33)}$$

(V-32) and (V-33) together give

$$\langle p|x\rangle = \langle x|p\rangle^*. \quad \text{(V-34)}$$

We emphasize that (V-34) does not follow from the "hermiticity property" of the "scalar product" $\langle x|p\rangle$ but is a very particular property of the operators P and Q (and -as a consequence thereof- of the Fourier transform)

For the general case of two arbitrary systems of generalized eigenvectors

$$|\lambda\rangle \; ; \quad A^{\times}|\lambda\rangle = \lambda\,|\lambda\rangle$$

and

$$|\beta\rangle \; ; \quad B^{\times}|\beta\rangle = \beta\,|\beta\rangle$$

it is <u>not</u> always true that

$$\langle\lambda|\beta\rangle = \langle\beta|\lambda\rangle^*.$$

though (1-67) holds always.

It is now simple to calculate the matrix element of P in the basis of generalized eigenvectors of Q, using (V-33):

$$\langle x|P|\varphi\rangle = \int dp\, p\, \langle x|p\rangle\langle p|x\rangle = \int dp\, (-i\hbar)\, d/dx\{\langle x|p\rangle\langle p|x\rangle$$

$$\langle x|P|\varphi\rangle = -id/dx\, \{\langle x|\varphi\rangle\} \quad \text{(V-35)}$$

In the same way, using (V-32), one obtains

$$\langle p|Q|\varphi\rangle = i\hbar\, d/dp\{\langle p|\varphi\rangle\} \quad \text{(V-36)}$$

It is now easy to see what the realization of Φ by the functions $\langle x|\varphi\rangle$ (or $\langle p|\varphi\rangle$) is. As every power Q^n and P^m of the operators Q and P are defined on Φ the functions $\langle x|\varphi\rangle = \varphi(x)$ must fulfill:

$$(\varphi, Q^n\varphi) = \int_{-\infty}^{\infty} dx\, x^n\, |\langle x|\varphi\rangle|^2 < \infty$$

$$(\varphi, P^m\varphi) = (-i\hbar)^m \int_{-\infty}^{\infty} dx\, \langle x|\varphi\rangle \frac{d^m}{dx^m} \langle x|\varphi\rangle < \infty$$

$$(\varphi, Q^n P^m \varphi) = (-i\hbar) \int_{-\infty}^{\infty} dx \, \langle x | \varphi \rangle \, x^n \frac{d^m}{dx^m} \langle x | \varphi \rangle < \infty$$

(V-37)

As all algebraic expressions in P and Q are continuous operators on Φ, the sequence $\varphi_n = \langle x | \varphi_n \rangle$, which is the realization of the convergent sequence $\varphi_n \rightarrow \varphi$ in the sense of τ_Φ, converges to $\varphi(x) = \langle x | \varphi \rangle$ if the $x^n \, (d^m/dx^m) \, \varphi_k(x)$ converge uniformly on every bounded region to $x^n (d^m/dx^m) \, \varphi(x)$

The space of functions $\varphi(x)$ that fulfill (V-37) and for which the convergence is defined in the above described way is the space of test functions (Schwartz space).

Concluding this appendix, we emphasize that in deriving (V-35) and (V-36) we have made use of (1-3), because by (1-3) we were led to (V-1) and therewith to the Hermite polynomials, whose properties have been used extensively to establish (V-35) and (V-36).

APPENDIX VI

PROOF OF PROPOSITION II IN CHAPTER IV

ε_t is a linear operator on S. S is a locally convex space in which the topology is given by the norms given in the expressions (1-110) to (1-113). As we can find in Reference 29 on Chapter I, $\hat{\varepsilon}_t$ is continuous on S if and only if for all n, m = 0,1,2,...

$$||E^n D^m \hat{\varepsilon}_t f(E)|| \leq \kappa\{ ||E^{p_1} D^{q_1} f(E)|| + ... + ||E^{p_k} D^{q_k} f(E)|| \}$$

(VI-1)

where κ is a positive constant, $D^m = (d^m/dE^m)$, the nonnegative entire numbers as well as κ depend on both n and m, and the norm here is the usual one on $L^2(R)$. We use the norms in (1-112) to check the continuity of $\hat{\varepsilon}_t$. Thus,

$$||E^n D^m \hat{\varepsilon}_t f(E)|| = ||E^n D^m e^{itE} f(E)|| = ||E^n\{(itE)^m e^{itE} f(E) + ... + e^{itE} D^m f(E)\}||$$

$$\leq |t|^m ||E^{n+m} f(E)|| + |t|^{m-1} m ||E^{n+m-1} Df(E)|| + ... + ||E^n D^m f(E)||$$

$$\leq (1 + |t|)^m \{ ||E^{n+m} f(E)|| + ||E^{n+m-1} Df(E)|| + ... + ||E^n D^m f(E)|| \} \qquad \text{(VI-2)}$$

which proves (VI-1).

APPENDIX VII
ON THE NONTRIVIALITY OF THE GAMOW VECTORS

The essential feature that the Gamow vectors $|Z_R^-\rangle$ and $|Z_R^{*+}\rangle$ must be nonzero, is equivalent to the assertion that the brackets $\langle\psi^+|Z_R^{*+}\rangle$ and $\langle\varphi^-|Z_R^-\rangle$ are nonzero for some $\psi^+ \in \Phi^+$ and $\varphi^- \in \Phi^-$. If all these brackets were zero our theory would be a nonsense, and the starting point equations (4-25) and (4-32) would be meaningless.

In order to prove that $\langle\varphi^-|Z_R^-\rangle$ cannot be zero for all $\varphi^- \in \Phi^-$, let us recall that the function of E, $\varphi(E) = \langle\varphi^-|E^-\rangle$, when E runs over R, belongs to Δ_-. All the functions in Δ_- are constructed in this way with φ^- running over Φ^-. Also:

$$\varphi(Z_R) = \langle\varphi^-|Z_R^-\rangle = -\frac{1}{2\pi i}\int_{-\infty}^{\infty} \frac{\varphi(E)}{(E - Z_R)}\, dE \qquad \text{(VII-1)}$$

is a consequence of the Titchmarsh theorem (Theorem I in Chapter II). Here, $\mathrm{Im}\, Z_R < 0$.

We need to prove that $\varphi(Z_R) \neq 0$ for some $\varphi \in \Delta_-$. Assume that $\varphi(Z_R)$ were zero for any $\varphi \in \Delta_-$. In such a case we could write (VII-1) as

$$\varphi(Z_R) = \frac{1}{2\pi i}\left(\frac{1}{E - Z_R^*}, \varphi(E)\right) = 0 \qquad \text{(VII-2)}$$

where the bracket denotes the scalar product on $L^2(R)$.

Since $h(E) = (2\pi i)^{-1}(E - Z_R^*)^{-1}$ belongs to H_-, its inverse Fourier transform $F^{-1}h(t)$ is in $L^2(R^+)$. Furthermore, when $\varphi(E)$ runs over Δ_-, $F^{-1}\varphi(t)$ runs over $S(R^+)$, which is dense in $L^2(R^+)$. From one of the properties of Fourier transform follows then:

$$\varphi(Z_R) = (F^{-1}h(t), F^{-1}\varphi(t)) = 0 \qquad \text{(VII-3)}$$

where the bracket in (VII-3) denotes the scalar product on $L^2(R^+)$. Since $S(R^+)$ is dense in $L^2(R^+)$, (VII-3) implies that $F^{-1}h(t) = 0$, which is nonsense because in this case $F\,F^{-1}h(t) = h(E) = 0$, due to the linearity of the Fourier transform.

Arguments of the same nature will prove that $\langle\psi^+|Z_R^{*+}\rangle$ cannot vanish for every $\varphi^+ \in \Phi^+$.

We conclude this appendix with the observation that not all the functions in Δ_- or in Δ_+ can be zero at the same poit $E \in R$. To fix our ideas assume that every function $\varphi \in \Delta_-$ vanishes at zero: $\varphi(0) = 0$.

If $\varphi(E)$ has this property, its inverse Fourier transform fulfills

$$\int_0^\infty F^{-1} h(t)\, dt = 0$$

(VII-4)

At this point, we state the following lemma: For every closed and bounded (compact) interval [a,b] in the real line, we can construct a real function, f(x), in S(R) such that: a) f(x) is positive and nonzero for $x \in (a, b)$; b) f(a) = f(b) = 0; c) f(x) is zero outside the interval [a,b]. (See for instance W.M. Boothby: "An introduction to differentiable manifolds and Riemannian Geometry", Academic Press, New York 1975, Chapter II).

Therefore, we can find functions in $S(R^+)$ such that their integrals cannot be zero, since tey are strictly positive over some interval contained in R^+ and are zero outside this interval. Their Fourier transforms are functions in Δ_- which do not vanish at zero.

Assume now that all the functions $\varphi(E)$ in Δ_-vanished at a nonzero real point, say α Then, we would have that

$$\int_0^\infty e^{-i\alpha t}\, \varphi(t)\, dt = 0$$

(VII-5)

for all $\varphi(t) \in S(R^+)$. For a real $\varphi(t)$, (VII-5) would imply that

$$\int_0^\infty \cos \alpha t\ \varphi(t)\, dt = 0$$

(VII-6)

and

$$\int_0^\infty \operatorname{sen} \alpha t\, \varphi(t)\, dt = 0$$

(VII-7)

According to the lemma stated above, we can construct a positive function f(t) in S(R) supported by the interval $[0, \pi/2\alpha]$ (f(t) is zero outside this interval). The integrals (VII-6) and (VII-7) do not vanish when we use f(t) for $\varphi(t)$, and, therefore, we reach to a nonsense.

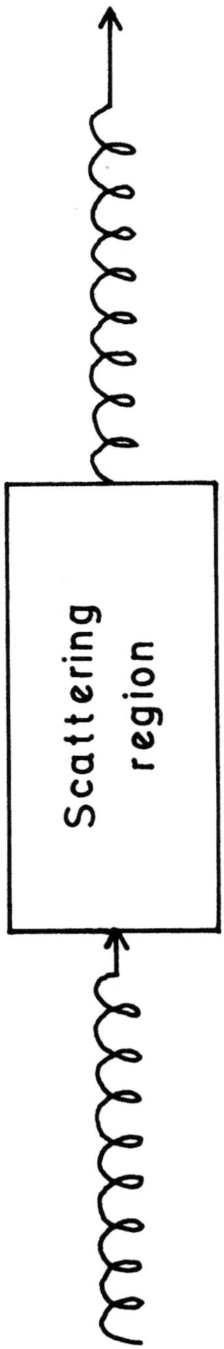

Figure 1

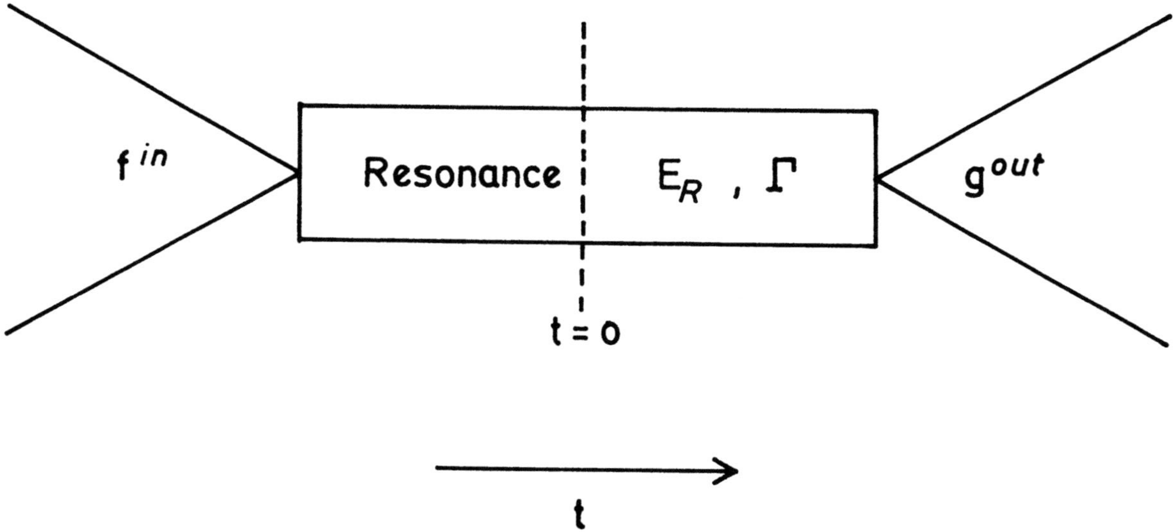

Figure 2

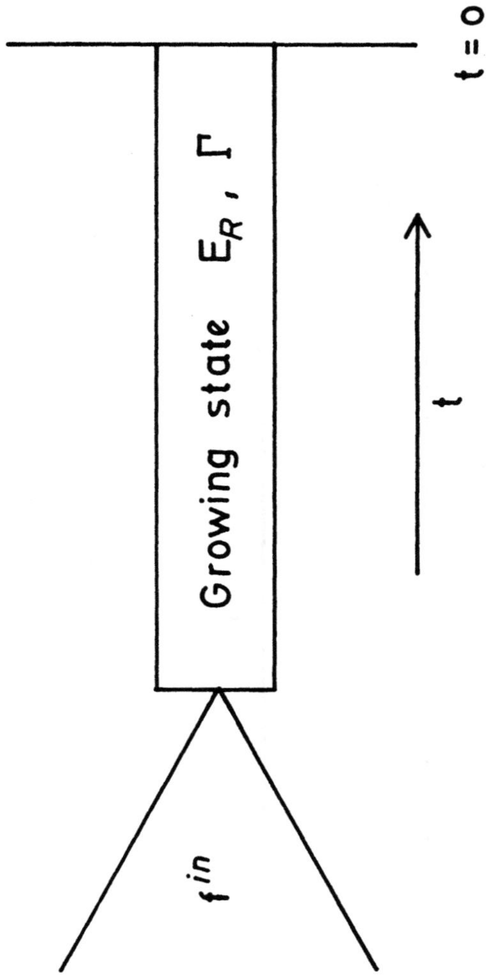

Figure 3

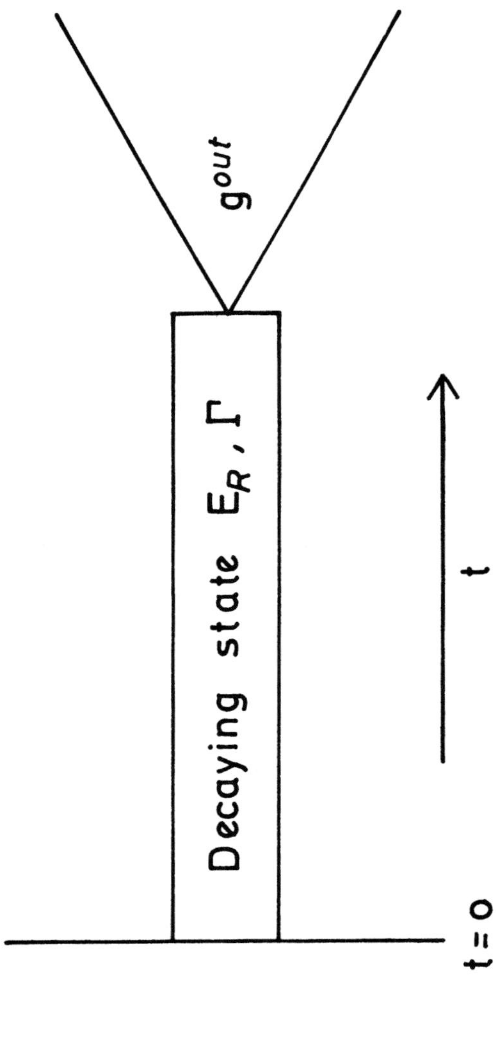

Figure 4

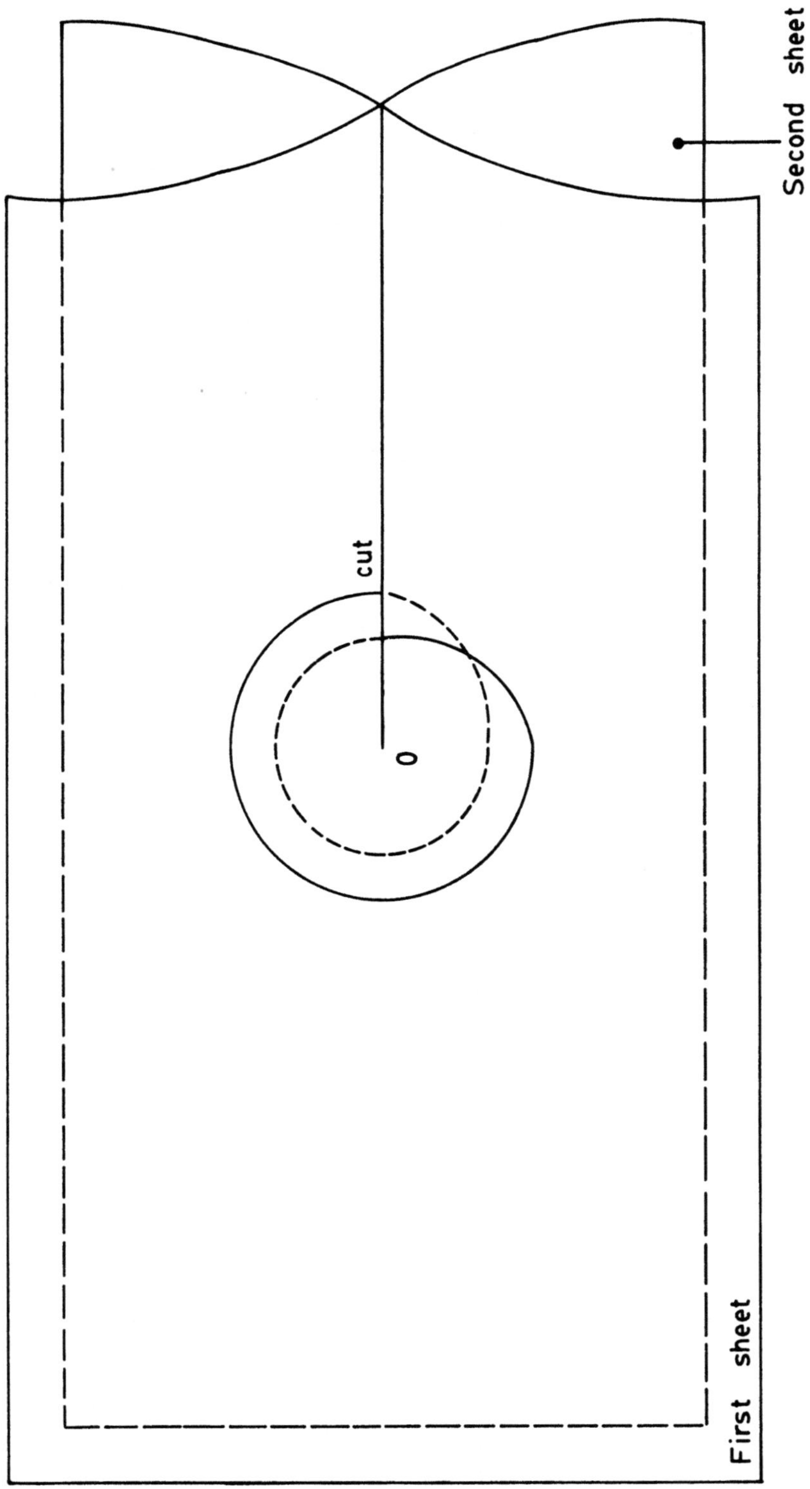

Figure 5

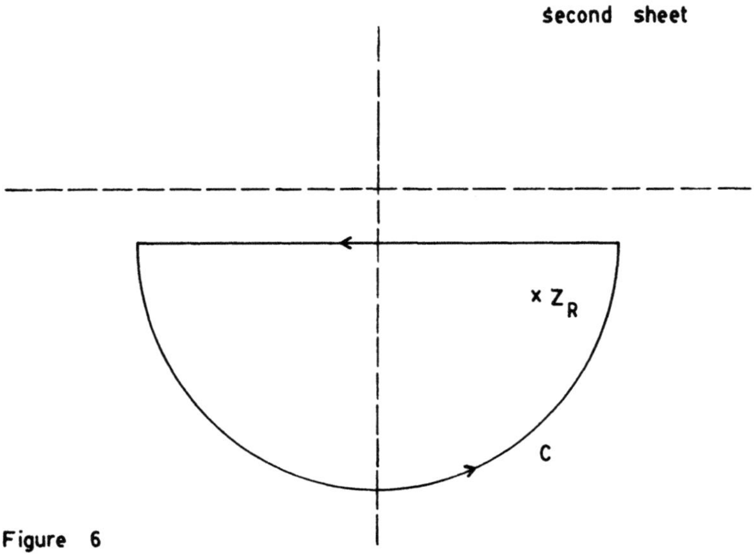

Figure 6

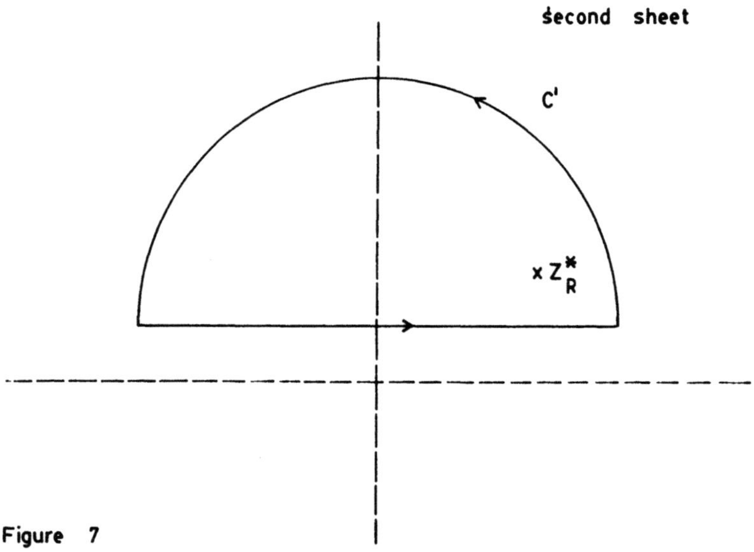

Figure 7

second sheet

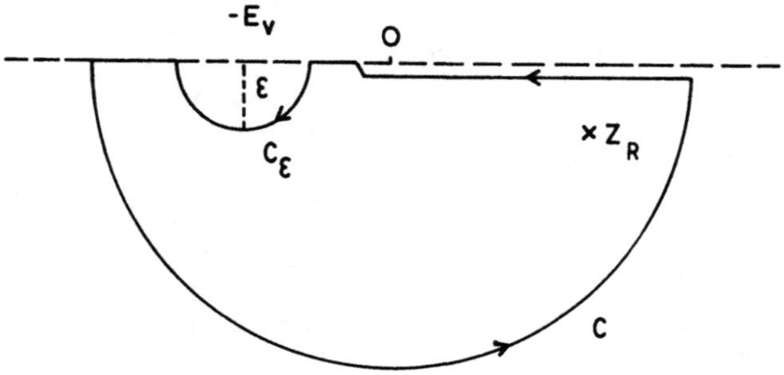

Figure 8

second sheet

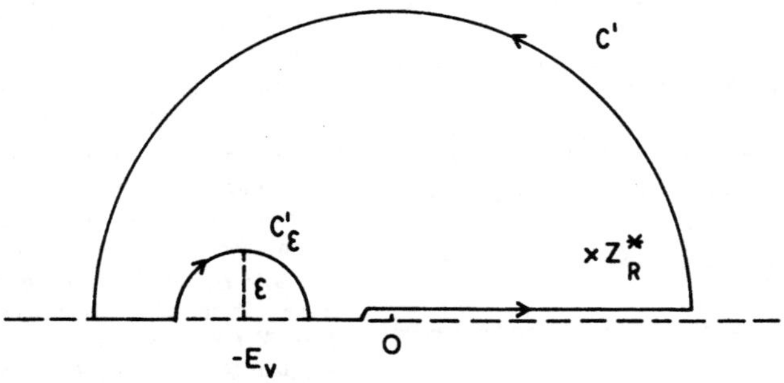

Figure 9

Lecture Notes in Mathematics

Vol. 1236: Stochastic Partial Differential Equations and Applications. Proceedings, 1985. Edited by G. Da Prato and L. Tubaro. V, 257 pages. 1987.

Vol. 1237: Rational Approximation and its Applications in Mathematics and Physics. Proceedings, 1985. Edited by J. Gilewicz, M. Pindor and W. Siemaszko. XII, 350 pages. 1987.

Vol. 1250: Stochastic Processes – Mathematics and Physics II. Proceedings 1985. Edited by S. Albeverio, Ph. Blanchard and L. Streit. VI, 359 pages. 1987.

Vol. 1251: Differential Geometric Methods in Mathematical Physics. Proceedings, 1985. Edited by P.L. García and A. Pérez-Rendón. VII, 300 pages. 1987.

Vol. 1255: Differential Geometry and Differential Equations. Proceedings, 1985. Edited by C. Gu, M. Berger and R.L. Bryant. XII, 243 pages. 1987.

Vol. 1256: Pseudo-Differential Operators. Proceedings, 1986. Edited by H.O. Cordes, B. Gramsch and H. Widom. X, 479 pages. 1987.

Vol. 1258: J. Weidmann, Spectral Theory of Ordinary Differential Operators. VI, 303 pages. 1987.

Vol. 1260: N.H. Pavel, Nonlinear Evolution Operators and Semigroups. VI, 285 pages. 1987.

Vol. 1263: V.L. Hansen (Ed.), Differential Geometry. Proceedings, 1985. XI, 288 pages. 1987.

Vol. 1265: W. Van Assche, Asymptotics for Orthogonal Polynomials. VI, 201 pages. 1987.

Vol. 1267: J. Lindenstrauss, V.D. Milman (Eds.), Geometrical Aspects of Functional Analysis. Seminar. VII, 212 pages. 1987.

Vol. 1269: M. Shiota, Nash Manifolds. VI, 223 pages. 1987.

Vol. 1270: C. Carasso, P.-A. Raviart, D. Serre (Eds.), Nonlinear Hyperbolic Problems. Proceedings, 1986. XV, 341 pages. 1987.

Vol. 1272: M.S. Livšic, L.L. Waksman, Commuting Nonselfadjoint Operators in Hilbert Space. III, 115 pages. 1987.

Vol. 1273: G.-M. Greuel, G. Trautmann (Eds.), Singularities, Representation of Algebras, and Vector Bundles. Proceedings, 1985. XIV, 383 pages. 1987.

Vol. 1275: C.A. Berenstein (Ed.), Complex Analysis I. Proceedings, 1985–86. XV, 331 pages. 1987.

Vol. 1276: C.A. Berenstein (Ed.), Complex Analysis II. Proceedings, 1985–86. IX, 320 pages. 1987.

Vol. 1277: C.A. Berenstein (Ed.), Complex Analysis III. Proceedings, 1985–86. X, 350 pages. 1987.

Vol. 1283: S. Mardešić, J. Segal (Eds.), Geometric Topology and Shape Theory. Proceedings, 1986. V, 261 pages. 1987.

Vol. 1285: I.W. Knowles, Y. Saitō (Eds.), Differential Equations and Mathematical Physics. Proceedings, 1986. XVI, 499 pages. 1987.

Vol. 1287: E.B. Saff (Ed.), Approximation Theory, Tampa. Proceedings, 1985–1986. V, 228 pages. 1987.

Vol. 1288: Yu. L. Rodin, Generalized Analytic Functions on Riemann Surfaces. V, 128 pages, 1987.

Vol. 1294: M. Queffélec, Substitution Dynamical Systems – Spectral Analysis. XIII, 240 pages. 1987.

Vol. 1299: S. Watanabe, Yu.V. Prokhorov (Eds.), Probability Theory and Mathematical Statistics. Proceedings, 1986. VIII, 589 pages. 1988.

Vol. 1300: G.B. Seligman, Constructions of Lie Algebras and their Modules. VI, 190 pages. 1988.

Vol. 1302: M. Cwikel, J. Peetre, Y. Sagher, H. Wallin (Eds.), Function Spaces and Applications. Proceedings, 1986. VI, 445 pages. 1988.

Vol. 1303: L. Accardi, W. von Waldenfels (Eds.), Quantum Probability and Applications III. Proceedings, 1987. VI, 373 pages. 1988.

Lecture Notes in Physics

Vol. 322: M. Schlichenmaier, An Introduction to Riemann Surfaces, Algebraic Curves and Moduli Spaces. XIII, 148 pages. 1989.

Vol. 323: D.L. Dwoyer, M.Y. Hussaini, R.G. Voigt (Eds.), 11th International Conference on Numerical Methods in Fluid Dynamics. XIII, 622 pages. 1989.

Vol. 324: P. Exner, P. Šeba (Eds.), Applications of Self-Adjoint Extensions in Quantum Physics. Proceedings, 1987. VIII, 273 pages. 1989.

Vol. 325: E. Brändas, N. Elander (Eds.), Resonances, Proceedings, 1987. XVIII, 564 pages. 1989.

Vol. 326: A. Grauel, Feldtheoretische Beschreibung der Thermodynamik für Grenzflächen. IX, 317 Seiten. 1989.

Vol. 327: K. Meisenheimer, H.-J. Röser (Eds.), Hot Spots in Extragalactic Radio Source. Proceedings, 1988, XII, 301 pages. 1989.

Vol. 328: G. Wegner (Ed.), White Dwarfs. Proceedings, 1988. XIV, 524 pages. 1989.

Vol. 329: A Heck, F. Murtagh (Eds.), Knowledge Based Systems in Astronomy. IV, 280 pages. 1989.

Vol. 330: J.M. Moran, J.N. Hewitt, K.Y. Lo (Eds.), Gravitational Lenses. Proceedings, 1988. XIV, 238 pages. 1989.

Vol. 331: G. Winnewisser, J.T. Armstrong (Eds.), The Physics and Chemistry of Interstellar Molecular Clouds mm and Sub-mm Observations in Astrophysics. Proceedings, 1988. XVIII, 463 pages. 1989.

Vol. 332: P. Flin, H.W. Duerbeck (Eds.), Morphological Cosmology. Proceedings, 1988. VII, 438 pages. 1989.

Vol. 333: I. Appenzeller, H.J. Habing, P. Léna (Eds.), Evolution of Galaxies – Astronomical Observations. Proceedings, 1988. X, 391 pages. 1989.

Vol. 334: L. Maraschi, T. Maccacaro, M.-H. Ulrich (Eds.), BL Lac Objects. XIII, 497 pages. 1989.

Vol. 335: A. Lakhtakia, V.K. Varadan, V.V. Varadan, Time-Harmonic Electromagnetic Fields in Chiral Media. VII, 121 pages. 1989.

Vol. 336: M. Müller, Consistent Classical Supergravity Theories. VI, 125 pages. 1989.

Vol. 337: A.P. Maclin, T.L. Gill, W.W. Zachary (Eds.), Magnetic Phenomena. Proceedings, 1988. VI, 142 pages. 1989.

Vol. 338: V. Privman, N.M. Švrakić, Directed Models of Polymers, Interfaces, and Clusters: Scaling and Finite-Size Properties. VI, 120 pages. 1989.

Vol. 339: F. Ehlotzky (Ed.), Fundamentals of Laser Interactions II. Proceedings, 1989. XI, 317 pages. 1989.

Vol. 340: M. Peshkin, A. Tonomura, The Aharonov-Bohm Effect. VI, 152 pages. 1989.

Vol. 341: E.F. Milone (Ed.), Infrared Extinction and Standardization. Proceedings, 1988. III, 79 pages. 1989.

Vol. 343: M. Month, S. Turner (Eds.), Frontiers of Particle Beams; Observation, Diagnosis and Correction. Proceedings, 1988. IX, 509 pages. 1989.

Vol. 344: M. Rascle, D. Serre, M. Slemrod (Eds.), PDEs and Continuum Models of Phase Transitions. Proceedings, 1988. VII, 229 pages. 1989.

Vol. 345: H. Holden, A. Jensen (Eds.), Schrödinger Operators. Proceedings, 1988. V, 458 pages. 1989.

Vol. 346: D. Lüst, S. Theisen, Lectures on String Theory. VII, 346 pages. 1989.

Vol. 348: A. Bohm, M. Gadella, Dirac Kets, Gamow Vectors and Gel'fand Triplets. VII, 119 pages. 1989.

GPSR Compliance

The European Union's (EU) General Product Safety Regulation (GPSR) is a set of rules that requires consumer products to be safe and our obligations to ensure this.

If you have any concerns about our products, you can contact us on ProductSafety@springernature.com

In case Publisher is established outside the EU, the EU authorized representative is:

Springer Nature Customer Service Center GmbH
Europaplatz 3
69115 Heidelberg, Germany

Batch number: 10683525

Printed by Printforce, the Netherlands